The Saga of the

U.S. Synthetic Fuels Corporation

By the same author

FREE PEOPLE, FREE MARKETS: Their Evolutionary Origins
(New Academia Publishing 2009)

Read an excerpt at www.newacademia.com

The Saga of the
U.S. Synthetic Fuels Corporation
A Cautionary Tale

RALPH L. BAYRER

NEW ACADEMIA PUBLISHING VELLUM

Washington, DC

Library of Congress Control Number: 2011937211
ISBN 978-0-9836899-3-5 paperback (alk. paper)

Contents

The Arab oil embargoes and energy upheavals of the 1970s had a dramatic impact on the American psyche, generating a fevered political response leading inter alia to the U.S. Synthetic Fuels Corporation.

The SFC mandate was to exploit America's vast oil shale and coal resources to provide alternatives to imported oil. There were many technologies that, based on pilot-scale testing, had the potential to convert those resources into such alternatives—if the technology and the economics could be proven at commercial scale.

After extensive hearings, consideration of many alternative approaches, and significant opposition, the Energy Security Act was passed by Congress in 1980.

The Energy Security Act set ambitious goals to establish a commercial synthetic fuels industry that could produce 2 million barrels a day within a decade. Its legislative provisions matched those ambitions with innovative financial instruments, Marshall Plan-scale funding of $88 billion, and the establishment of a unique quasi-federal entity, the SFC.

Acknowledgements

This book is dedicated to Edward E. Noble, who was Chairman of the U.S. Synthetic Fuels Corporation for most of its existence. His profound instincts in the ways of energy markets, along with those of his fellow board members, were instrumental in reshaping an overly ambitious government program into one that would meet national strategic energy needs at modest cost and with minimal risk to the taxpayer. His grace under a political fire that is common to contentious issues in the Washington environment was an inspiration to me and much of the SFC staff.

In completing this work, I am indebted to the assistance of individuals who were exceedingly generous with their time and critical analysis. Doug Seay and Dr. Jack Prostko helped make the narrative clearer, more balanced, and more appealing to a general audience that might be as interested in the energy crisis and political climate surrounding the synthetic fuels program as in the particular history of the Corporation.

And, of course, the loving support of my life partner Ken George was essential to my being able to bring this work to fruition.

Introduction

The Saga of the United States Synthetic Fuels Corporation: A Cautionary Tale presents a history of the U.S. Synthetic Fuels Corporation (SFC) set in the context of the country's last energy crisis, which hit in the 1970s. Although Congress launched the Corporation as a major national initiative, authorizing it $88 billion ($230 billion in 2010 dollars)—twice the size of the Marshall Plan that rebuild Europe after the Second World War)—its history and the results of its endeavors are barely known today. To be sure, the final outcome was far more modest than the "moral equivalent of war" claimed by President Carter. But much was learned that can inform the country as it grapples anew with energy policy. It is also an untold story of political conflict in the nation's capital, a cautionary tale about grand energy crusades.

The SFC was created when the country had grown anxious about an apparent loss of energy security, as demonstrated by skyrocketing prices, a disconcertingly sudden dependence on foreign oil, and a sense of vulnerability induced by foreign countries' willingness to use oil embargoes as weapons. This book shows how Congress and the president reacted to these circumstances. Congress initially passed some modest legislation strengthening conservation efforts and accelerating research for alternative fuels. But after the Iranian Revolution of 1979 and another bout of accelerating oil prices, President Carter determined that a much more dramatic national effort was required. The SFC would be that effort's centerpiece.

Synthetic fuels, alternatives to petroleum and natural gas derived from solid resources, were the focus of the new effort. The

United States' vast non-petroleum energy resources, largely coal and oil shale, contain the energy equivalent of all the oil in the Middle East many times over. Past efforts in other countries appeared to endorse endeavors to employ new technology to address strategic insecurity, notably Germany's synthetic fuels industry that was created when the Second World War shut off that country's petroleum supplies and South Africa's undertaking a synthetic fuels industry that met more than half that country's petroleum needs when it appeared that Arab embargoes might shut it out of petroleum markets.

Although many technological approaches for producing liquids and gases from these solid resources had been identified and attempted at a pilot scale in the United States, they had not been proven at commercial scale. Doing so would entail building plants ten or more times larger than had ever been attempted and which would have cost billions of dollars apiece. Moreover, experience in other industries showed how difficult unanticipated operating problems could be. Companies simply would not take on all of those risks themselves. Thus, in 1980 Congress created the SFC, a quasi-federal institution, to assist the private sector in creating an entirely new industry, one that was to produce 2 million barrels a day of synthetic fuels within a decade, thereby displacing a significant fraction of imported oil.

The book recounts the history of the six-year life of the SFC: its structure and operating philosophy, how it solicited and evaluated proposals from the private sector, how it negotiated assistance contracts, as well as specific aspects of the synthetic fuels projects it supported. In so doing, the book shows how changing energy circumstances and the realities of the market place inexorably moved the Corporation away from the grand aspirations of the Energy Security Act to a more sensible development of a modest array of key technologies at commercial scale, which the country would have at the ready whenever they should be needed.

As it happened, in 1985—a mere five years after it launched its ambitious effort to achieve energy security—Congress fecklessly terminated the Corporation, thereby aborting both the grand effort as well as the course correction that it and the SFC's management belatedly determined to be economically and strategically sensible.

Although the SFC had completed financial negotiations with about thirteen projects that represented a promising diversity of energy resources and technologies, the congressional action terminated the Corporation after assistance contracts had been finalized with but four projects.

The book summarizes the history of the four projects the SFC funded, as well as that of another funded by the Department of Energy that was originally intended for transfer to the SFC. It relates the extensive experience gleaned regarding how well these technologies work at full scale, what they cost to build, how quickly experience reduced operating costs, and just how difficult debugging new technologies can prove to be. It also summarizes their environmental performance, which had been intensively monitored.

Finally, the book presents an illuminating case study of how a determined ideological constituency can work its way in the national political arena in the face of what appears to be a general consensus, and in this instance contrary to the staunch support for the Corporation by the leaders of both parties in both the House of Representatives and the Senate. Ultimately, the outcome can be traced to a fall in oil imports (however temporary), a decline in gasoline prices (also temporary), and waning public interest. But in the instance of the SFC, unremitting hostility by the environmental community, hostile congressional hearings, repeated GAO investigations (which produced no findings to the detriment of the SFC), and demagogic speeches on the floor of congress over the years took their toll as well, leading to the SFC's untimely end. Now merely a few decades later, virtually nothing is known of this—no books have been written and other media are seemingly oblivious to the program's accomplishments.

Other than a filling a bewildering gap in the historic narrative, what does this book offer? It offers necessary perspective at a critical juncture. The SFC experience can provide pragmatic lessons for structuring future governmental programs in partnership with the private sector (if any) and provide cautionary lessons for Congress to keep in mind before it contemplates any new energy crusades. Five pioneer commercial-scale production facilities (one under the aegis of the Department of Energy before the SFC became operational) were built and operated. These can provide operating experience essential for exploiting some of America's vast energy resources that are in solid rather than liquid or gaseous forms, whenever and however that might be justifiable on economic or national security grounds. In addition, years of operating experience by these projects demonstrated that these new technologies were environmentally benign with regard to air, liquid, or solid emissions and produced either no offsite discharge or operated comfortably within regulatory limits. Because prior to the program concerns existed about the potential environmental impact of these new technologies, the Energy Security Act required project sponsors to develop detailed environmental monitoring plans. Accordingly, these projects undertook years of monitoring under the guidance of the Corporation, the Department of Energy, and the Environmental Protection Agency. This experience is also now available to any that might need it in future energy developments.

Moreover, the Corporation's experience with the innovative financial methods authorized by the Energy Security Act demonstrated the utility of price and loan guarantees to encourage the private sector to build sizeable facilities with surprisingly modest government support. These had the added virtue of keeping the technological and commercial design of these facilities under the management of those who understood commercial needs the best. Despite uninformed claims in relatively recent media articles that the SFC wasted billions of dollars, the costs to the government of all four commercial-scale facilities and all the administrative costs of the Corporation came to less than one billion dollars.

Negative lessons to be drawn from experience during the 1970s generally and with regard to the ambitions of the Energy Security Act in particular show how the federal government, president and Congress alike, tends to overreact and reach for grandiose solutions. The book shows how some crises were self-inflicted by governmental distrust of free markets. For example, the extensive price controls, regulations, and excess profit taxes enacted in the U.S. created shortages and a sense of panic experienced by no other developed country, even those with no domestic production of petroleum whatsoever. Moreover, proposed solutions were too often driven by dramatic rather than pragmatic and prudent considerations. Notably, during the congressional debate over the

Energy Security Act, there were a number of proposals to carry out a limited program to first prove technologies before undertaking a massive production effort. These lost out to desires to create a major new industry in short order commensurate to the perceived national security challenge. Eventually, market reality forced the Corporation and a later Congress to accept the wisdom of a more limited approach, before Congress lost its appetite for synthetic fuels entirely. Ultimately, the government's credibility suffered. Congress appeared feckless, private corporations became more distrustful of doing business with the government, and the concept of a quasi-federal corporation was unfairly discredited.

Today, the United States has once again reached a political juncture in which the government is overdramatizing a perceived problem—in this case, the threat of global warming. In response, Washington is attempting to impose a vast new regulatory structure on energy sectors of the economy and to spend substantial monies on still uneconomic 'green energy.' The experience of President Carter's moral equivalent of war suggests that economic reality will ultimately corral the more grandiose initiatives being considered in present day debate, but perhaps not before they punish the U.S. economy with costs far beyond those of the synthetic fuels program.

The following fourteen chapters are intended to help complete the sorely deficient historical record by portraying how a past energy crusade unfolded, thereby shedding light on current energy policy debates.

I

The Political Impetus

Today, renewed concern about the availability of energy resources is provoking new interest in lessons learned from past "crises," in particular the energy market upheavals of the 1970s. Those crises were characterized by growing U.S. dependence on imported oil, oil embargoes, wars in the Middle East, and soaring prices for energy across the board. Political responses under the banners of "energy independence" and the "moral equivalent of war" went in many directions: the imposition of widespread controls on the economy, the creation of Department of Energy, investment by the government in energy research and development, and an attempt to create a commercial synthetic fuels industry led by the U.S. Synthetic Fuels Corporation (SFC). The political context, the origins, the structure, and the life of the SFC provide a particularly useful prism for viewing that energy era and for extracting lessons regarding what the energy future might have in store for us.

To that end, the book provides an overview of the country's vast coal and shale resources, of the readiness of likely technologies for drawing upon these, and of various financial instruments available to enlist the private sector in developing pioneer technologies. In that context, it lays out the political process that led to the passage of the historically ambitious Energy Security Act, which aimed at creating a new energy sector in the economy. It recounts the history of the six-year life of the SFC: its structure and operating philosophy, how it solicited and evaluated proposals from the private sector, and how it negotiated assistance contracts, as well as the operating and environmental experience gleaned from the projects it supported. In so doing, the book shows how changing energy circumstances and the realities of the market place altered the political temper of Congress to first modify the SFC's mandate and then to terminate it prematurely. These experiences shed light on key factors for the success of new technology development and on how congressional intentions must be congruent with economic reality to succeed.

This chapter begins with an examination of the political context of the energy upheavals of the seventies to show why they had the dramatic impact on the American political psyche that they did, and concludes with a review of the U.S. political response.

ENERGY: FROM SELF-SUFFICIENCY TO DEPENDENCE ON OTHERS

Modern affluence depends utterly on the economic availability and utilization of energy. Illumination, heating and cooling, mobility, agricultural processes, and the production processes of modern industry would shrivel without it. If energy security were only a matter of units of energy, the United States would have no reason ever to be concerned about energy shortages. As will be shown in Chapter II, the United States is arguably the best endowed country in the world with regard to energy resources, considering petroleum, natural gas, coal, oil shale, and uranium. Different forms of energy, however, are not fungible. While a considerable substitutability exists among the abundant resources for heating or for generating electricity, many forms of transportation uniquely require some form of liquid fuels (electric cars or fuel cells notwithstanding). This is true for automobiles, trucks, aircraft, ships, and virtually all military vehicles. Consequently, without adequate liquid fuels, our transportation infrastructure, our military preparedness, and our personal mobility would be greatly impaired.

It was the great fortune of the United States for much of the previous century to possess ample supplies of petroleum. Nonetheless, because of the inherent uncertainties in discovering and exploiting new fields, and given the importance of petroleum to the country, circumstances periodically did give rise to concern about the Unit-

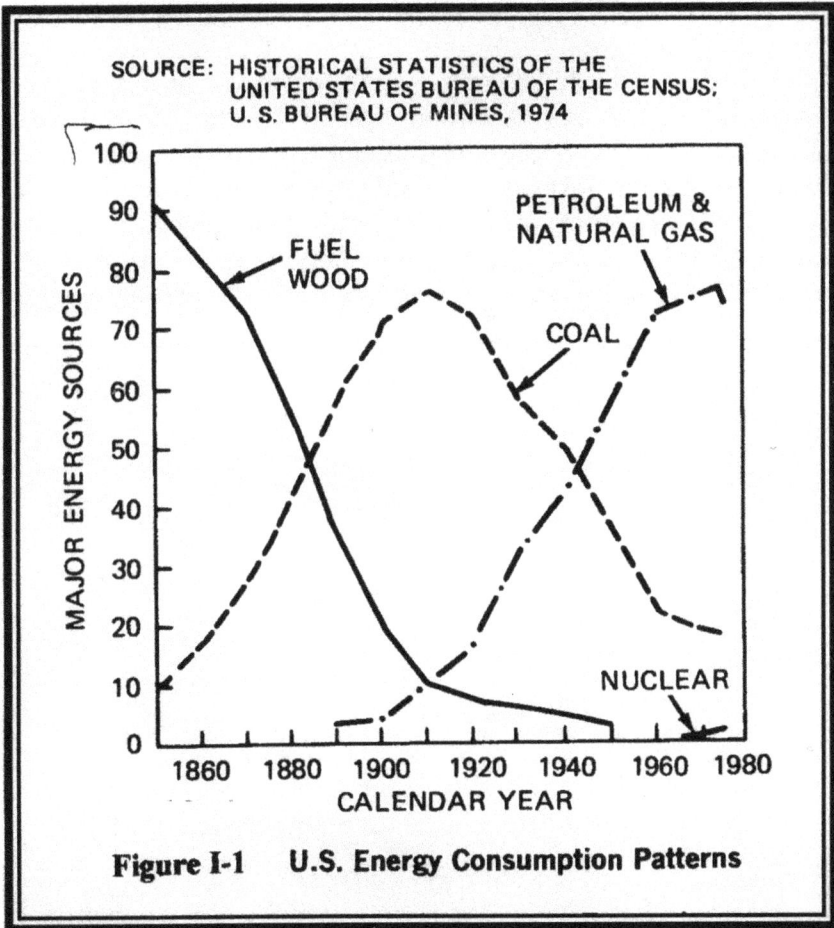

SOURCE: HISTORICAL STATISTICS OF THE
UNITED STATES BUREAU OF THE CENSUS;
U. S. BUREAU OF MINES, 1974

Figure I-1   U.S. Energy Consumption Patterns

ed States being able to meet its needs in future times. A brief history of the ebb and flow of such concern is useful in understanding the country's reaction to the perceived energy crises of the 1970s.

Early Self Sufficiency

Figure I-1, which depicts the contributions of various energy sources to meeting U.S. demand over a 150-year period, shows how petroleum and natural gas displaced wood and coal as the primary sources of energy.[1] By 1973, at the time of the Arab oil embargo, the

United States met 75 percent of its energy needs from petroleum and natural gas.

Oil had been found seeping from the earth and in salt wells for some decades, but in small quantities. In these amounts, oil did not attract any interest as a substitute for wood and coal because it was too expensive. Instead, its initial importance was as illuminating fuel and as a source of lubricant for industry because the vegetable and animal fats being used were too limited in quantity and too expensive. All of this changed when entrepreneurs deployed

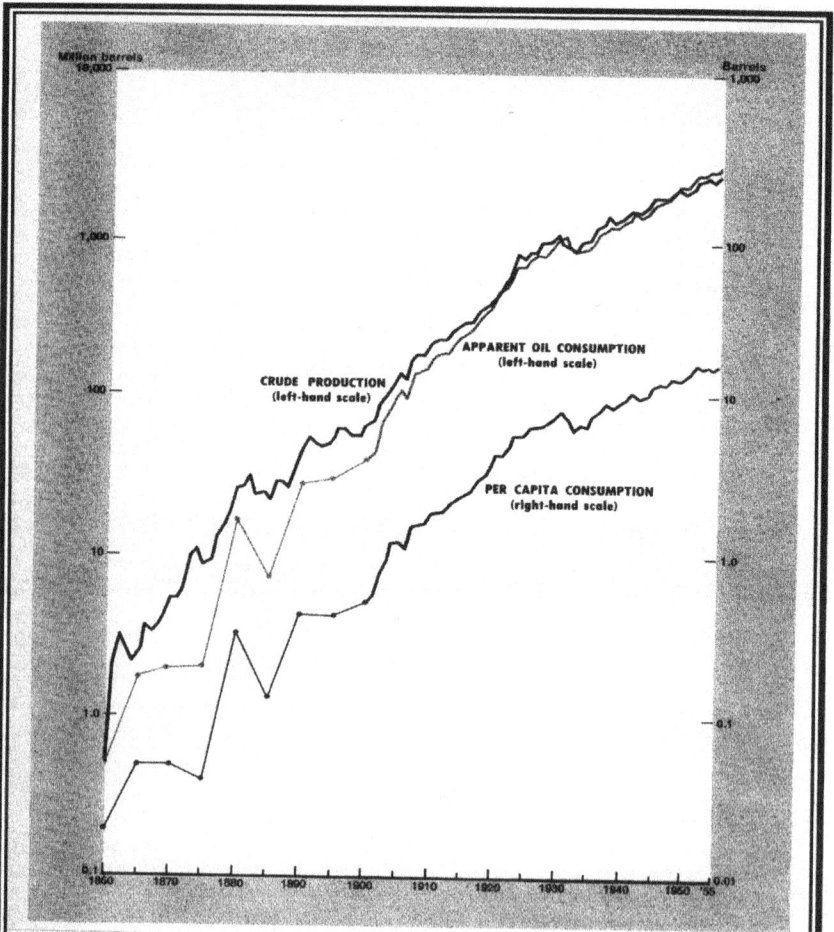

Figure I-2   *Production and consumption of oil, 1860–1955.*

rudimentary drilling rigs developed for salt mines to seek oil in Pennsylvania. On August 27, 1859, the first successful oil well was drilled at Titusville, Pennsylvania.[2] Production and use grew rapidly as new applications for petroleum were discovered, as seen in Figure I-2, Production and Consumption of Oil, 1860-1955, which illustrates the explosive growth.[3] Petroleum did not begin to substitute for coal until larger and lower production cost fields were found in Lima, Indiana.

Petroleum's greatest use-advantage became apparent as a mobility fuel when its newfound availability made feasible wide-scale use of existing technologies. For example, the concept of internal combustion engines had been theoretically understood but not aggressively pursued until the sources of petroleum were found. In 1876, the first gasoline engine was constructed in Germany, and in the early 1890s the first motor cars were developed. (Steam and electric propelled vehicles had existed before this time.) The growth of gasoline consumption was at first slow because existing roads were inadequate to support the rapid adoption of automobiles. But as the quality of the technology and the road system improved, the value of automobiles for mobility in our vast land led to the rapid growth of a new industry.

Petroleum's superior utility for warships also quickly became apparent. Notably, Winston Churchill, as First Lord of the Admiralty, ordered the conversion of the English war fleet to oil. Because he was aware of the consequences of supply vulnerability that England would be incurring in the attempt to achieve greater speeds and range, he convinced the House of Commons to invest in the Anglo-Persian Oil Company to ensure reserves in the event of war.[4] Planners similarly soon realized that air travel would be impossible without fuels derived from petroleum, and the growing mechanization of ground combat increased armies' reliance on the fuels as well. Accordingly, during the First World War the U.S. for the first time consumed more oil than it produced. Such imports (largely from Mexico) continued into the 1920s.

These first signs of dependence spurred Congress to several actions. It passed the depletion allowance tax provision to provide additional incentives for the domestic industry to explore for new reserves and to develop them. In addition, it opened public lands for exploration. Moreover, it encouraged American firms to explore abroad (which they eventually did with spectacular success in the Middle East). These measures along with natural enterprise led to a disappearance of the fears of shortages: domestic production of oil doubled from 1918 to 1923, and production continued to grow through the 1920s. When the enormous East Texas oil field was discovered in 1930 and the Depression reduced the domestic demand for petroleum, the nation's policy focus ironically moved to issues associated with controlling *overproduction*.

The great demands of the Second World War, largely met through domestic sources, and of the ensuing post-war boom again drew the attention of the federal government to the question of ensuring the sufficiency of future petroleum supplies. In the early fifties, the President's Materials Policy Commission, chaired by CBS head William Paley, alerted the nation to the limited nature of its petroleum reserves. In response, the government supported the development of a number of synthetic fuel technologies at the pilot plant stage. But the time for synthetic fuels had not yet arrived. Such technologies were infeasible given the large amounts of inexpensive oil available from abroad—in the Middle East, Nigeria, Mexico, and Venezuela. And because extracting oil from these sources was far less expensive than expanding production at home, domestic production was displaced by imports. Indeed, production from the Middle East and North Africa was growing by leaps and bounds, causing the real price of oil to decline.

The stage was being set for growing dependence on foreign oil and, inevitably, energy vulnerability. Never again would the United States be self sufficient in petroleum: Figure I-3, Annual U. S. Crude Oil Imports from All Countries, shows that petroleum imports grew steadily in the fifties and sixties, and alarmingly in the seventies.[5]

The first sign of energy insecurity becoming a factor in geopolitical events came in 1956 with the outbreak of the Arab-Israeli War, when Britain and France tried to wrest control of the Suez Canal back from the Egyptian leader Gamal Abdel Nasser. Because at this time Europe was getting about three-quarters of its oil from the Middle East through the Suez Canal, Egypt sank ships in the Canal to close it for six months. Europe would have

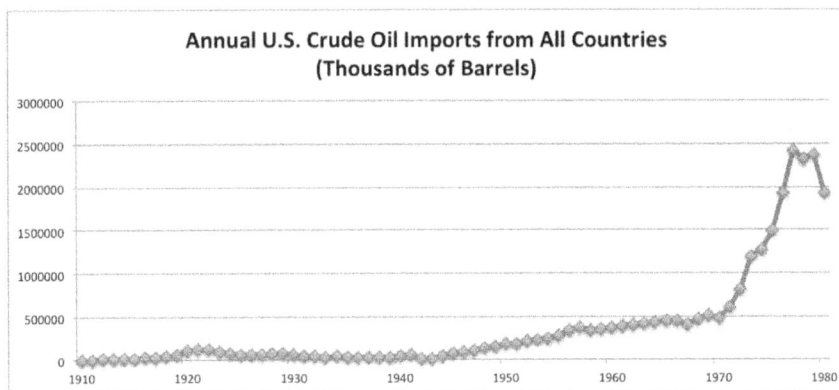

Annual U.S. Crude Oil Imports from All Countries
(Thousands of Barrels)

Figure I-3

had a desperate winter without the United States, which used all of its spare capacity to boost production by 2 million barrels a day to meet Europe's needs.[6] The United States, however, would never again be in a situation where it could meet energy shortfalls elsewhere. Recognizing the likelihood that Arabs would use U.S. oil dependency as a political weapon again, in 1959 President Eisenhower tried to curtail the growing levels of oil imports by the U.S. by imposing import restrictions even though domestic oil was more expensive to produce than foreign oil.

Changing Energy Fortunes

The vast wealth represented by petroleum reserves led to attempts by some producing countries to expropriate the Western companies that had originally explored, developed, and managed those fields. The first move came in 1960, when the Organization of Petroleum Exporting Countries (OPEC) was formed under the leadership of Saudi Arabia and Venezuela. This move was not taken very seriously at the time, given that the world's major producers—the United States and the Soviet Union—were not a part of this attempted cartel.

But events were to move in the direction of enhancing OPEC's role, not just because of the economic opportunity but because of the possibility of using control of petroleum resources as a geopolitical weapon. In particular, the share of world oil production from the Middle East, a region of potential instability and festering sense of grievance in the Arab states regarding Israel, continued to grow. Indeed, this political instability led in due course to another war between Israel and its Arab neighbors in 1967. Because of the war's brevity and Israel's unambiguous victory, there were no significant disturbances in the world petroleum markets. The world would not, however, be spared economic shocks the next time around.

Growing Sense of Vulnerability

Until the 1970s, the public remained generally unaware of the nation's increasing vulnerability. Indeed, in 1969, Congress, despite a declining share of domestic production in meeting the nation's needs, in a fit of anti-corporatism, cut the depletion allowance from 27.5 percent to 22 percent. This had a predictable impact on the independent oil companies, which carried out the bulk of the domestic exploration drilling. Such drilling declined rapidly, so that in 1971 the number of wildcats fell to their lowest level since 1947.[7]

While the year 1970 brought energy issues more fully into the public awareness, unfortunately, the public believed that incipient energy shortages were the artificial result of market manipulation by the large oil companies. Never mind that problems were worldwide. Never mind that study after study by Congress and others emphasized that the problems resulted from market forces. The majority of the American people were determined to believe that energy problems resulted from conspiracy rather than politics and markets. There are obvious parallels with the current political climate. This disconnectedness of the electorate from energy reality undoubtedly accounts for why some of the political actions worked at cross-purposes to their stated goals, as discussed below.

In addition to reducing industry's incentive to explore for oil, Congress contributed to shortages in several ways. It failed to remove price controls from natural gas, thereby reducing incentives to develop new sources and to ship gas, helping to ensure interstate shortages. In 1967, it had passed the Clean Air Act, which placed restrictions on the emissions of sulfur oxides from burning coal, thereby inhibiting coal's ability to substitute for petroleum where technology permitted. In 1970, it passed stiffer restrictions still. Because utility plant scrubbing technologies were not yet economically viable, the utilities shifted much of their generating capacity to oil-burning facilities that drew heavily on North African low sulfur oils. Consequently, generating capacity tightened and utilities became more dependent on oil. One result was that the East Coast experienced brownouts during the 1969 heat waves, and more were expected during the winter of 1970. Then, in May 1970, the Tapline, a pipeline crossing Syria that carried half a million barrels of oil daily to the Mediterranean, was accidentally severed. The Syrian government opportunistically refused to allow repairs to be made until higher transit fees were agreed to, effectively closing the line for nine months. During the outage, more oil had to be brought by tanker around Africa, resulting in a worldwide tanker shortage. Given this opportunity, the Libyan government began cutting production back by 600,000 barrels a day to gain leverage over the oil companies so as to force them to pay higher prices and taxes to Libya.

Political antennae went up and Congress did what it always does to deflect accountability. It held hearings. President Nixon appointed a White House committee to recommend policies to ensure an adequate supply of energy for the next five years. The secretary of the Interior appointed a committee to study the projected supply situation between the years 1985 and 2000. Yet, despite the fact that oil consumption was growing rapidly elsewhere in the world, no shortages occurred during the winter of 1970-71—thanks largely to industry being able to draw upon reserve crude production capacity. When the Tapline was reopened in January, prices declined and the sense of urgency temporarily waned.

Price and Profit Concessions

But with the growing evidence of the vulnerability of consuming nations, the Middle Eastern and North African nations increasingly forced concessions from the international oil companies who had developed those fields. Eventually, production was largely nationalized in a number of countries having the most reserves of petroleum. These events were triggered when Moammar Qaddafi seized power in Libya in 1969 and extended by the Shah of Iran and the leaders of other OPEC countries when the economic opportunities became obvious. Four months after the coup that brought Qaddafi to power, the twenty producing companies were summoned to a meeting.[8] The government announced that demands would be made for higher crude oil prices. Qaddafi was reported as stating that "People who have lived for five thousand years without petroleum are able to live without it even for scores of years in order to reach their legitimate right."[9] Indeed, given Libya's small population of 2 million and ample hard currency reserves, they were in a position where they could easily force the oil companies to reduce their production and revenues without seriously affecting the well being of the Libyan people.

Libya began with the most vulnerable company, Occidental Petroleum. It reduced allowed production levels until the company made generous concessions. Although Occidental was the initial target, pressure was eventually brought to bear on most of the companies. At this time, Libya exported about 3 million barrels of oil a day, making it the third largest oil-exporting nation in the world. By the time its campaign was completed, production had dropped to 800,000 barrels per day. In response, the United States activated its Emergency Petroleum Supply Committee and warned that it no longer had the means to assist other nations as it had in 1967.[10] When the smoke cleared, Libya taxed profits of the companies at rates about 55 percent, and the prices for the oil had been raised. In response, three companies, Gulf Oil, Atlantic Richfield, and Grace Petroleum gave up their oil production concessions.

Other nations in the Middle East were poised to negotiate comparable deals. Iran demanded the same tax rates, but the issue of oil prices was delayed until an OPEC meeting that was scheduled for December 1970, in Caracas, Venezuela. Beforehand, Venezuela raised the baseline. Even though Venezuela already had the highest taxes on profits of companies of any of the oil exporting nations, the Venezuelan congress bumped the tax rates further, so that the government would receive eighty percent of the profits from oil production overall.

At this point, all of OPEC got in the act to renegotiate prices with the oil companies in January. OPEC threatened "concerted and simultaneous action by all member countries" to enforce their objectives if they were not achieved in approximately three weeks of negotiation.[11] Libya then decided to upstage the rest of OPEC (despite having just concluded negotiations in its country) by presenting a new set of demands—which were three times as costly as the demands negotiated just three months earlier.

U.S. companies tried to present a united front by getting a ruling from the Justice Department that they could negotiate as a group without worrying about anti-trust action. The State Department got involved as well. But given the complexity presented by a large number of companies worldwide dealing with multiple countries, the united front approach eventually collapsed and the Gulf countries' demands were met. They received an immediate 35-cent per barrel increase in posted prices, to be followed by additional increases in June 1971 and January of each of the following three years. In addition, they would receive increases of 2.5 percent on the same dates to compensate for inflation. Finally, they all agreed to a fixed tax on profits of 55 percent.[12]

So while major producing companies had previously achieved a balance of petroleum supply and demand resulting in price stability, the actions of producing countries led to much higher prices and greater market instability.

Nationalization

Having extracted price hikes and profit share increases so easily, producing countries moved on to bigger game: the nationalization of the companies that had developed the fields. In December 1971, Libya nationalized the interests held by British Petroleum. This action was largely motivated by political rather than economic considerations when Qaddafi became enraged that the British had not prevented Iran from a military takeover of three islands in the mouth of the Persian Gulf.[13]

Iran moved next in 1973 when it concluded an agreement in which Iran would have 100 percent ownership of all oil and facilities. The oil companies were bumped from the status of owners to that of customers with the preferred right to buy and distribute Iran's oil. During the negotiations that led to this arrangement, Iraq nationalized its northern fields run by the Iraq Petroleum Company, while Libya dispensed with negotiations and delivered an ultimatum to the oil companies. They were to receive only net book value for the takeover and get to sell the oil at market prices, with no guarantees by Libya.[14]

Clearly, a new order was emerging. The uncertainty accompanying its birth was creating stress throughout the consuming world, along with the demoralization associated with a decline in the consuming countries' relative power. Nonetheless, the manner in which producing nations claimed control of their own resources and assumed control of world oil prices need not have been so traumatic. It was the attempt to use the West's growing dependency on imported oil as a political weapon that shook the political consciousness of the United States.

OIL AS A POLITICAL WEAPON

As mentioned earlier, Arab nations had used oil as a political weapon before the 1970s but without any dramatic impact because of the surplus capacity of the United States and the production of non-Arab nations, which blunted the Arab nations' oil weapon. A serious threat would require some sort of collective action by producing states. Such a threat emerged with the onset of the Yom Kippur War in October 1973—the fourth Arab-Israeli conflict in 24 years. Ministers of eleven members of the Arab Petroleum Exporting Countries agreed to cut exports to force a change in the United States Middle East policy. The next day, Saudi Arabia announced that it was slashing oil production and would cut off all shipments to the United States if it continued to supply arms to Israel and refused to modify its pro-Israel policy.[15]

By the second week of the war, the Arabs had reduced their production of oil by about one-quarter and embargoed shipments to the United States and the Netherlands, initially, and then to Portugal, South Africa and Rhodesia as well. By this time, the United States was importing about 37 percent of its oil needs, although only 5 percent of total consumption was coming from Arab sources. But Arab oil was meeting 73 percent of Western Europe's needs and 45 percent of Japan's, and so the Arab governments hoped also to pressure the United States through its allies.

This action should not have come as a surprise. In April 1973, King Faisal of Saudi Arabia sent his oil minister, Sheikh Yamani, to Washington with a message that Saudi Arabia would find it difficult to increase its oil production for export to the United States if the U.S. did not help to settle the Middle East problems to the satisfaction of the Arab states.[16]

With these events, the American populace was made dramatically aware of the nation's vulnerability to oil interruptions, even though they were rarely clear as to the reasons. A number of examples will suffice to demonstrate the wide-ranging nature of the impact on society. *Energy Crisis: Volume 1, 1963-73* provides news summaries of these eventful years:

Testimony to the Senate Interior Committee on October 24 estimated the petroleum shortfall to the United States at 2 million barrels per day or twelve percent of the total 17 million barrels a day used by the United States.

The nation's defense forces were experiencing energy shortfalls. According to an article in the *Washington Post* on October 26, the fuel needed for ships and planes fell short by about 10 percent. At this time the Pentagon purchased about 20 percent of its oil from Middle East sources.

Crude oil prices quadrupled, from about $3 per barrel to about $12 per barrel, with commensurate increases in refined products like gasoline.

Long gasoline lines appeared, with waits on the order of hours. Often a station would be sold out before the line of customers could be serviced. Conditions were particularly stressful over the Christmas and New Year holidays when many stations were closed because they either followed the voluntary Sunday sales ban or they had exhausted their monthly gasoline deliveries.

Independent truckers launched a four-day protest from December 4 to December 7 to protest rising fuel costs and the government-ordered reduction in speed limits. Traffic was halted by "stall-ins" at major thoroughfare bottlenecks. On December 13, they began a two-day strike marred by violence, which included brick throwing and 35 incidents of gunfire.

The big airlines made spot cancellations in service to conserve fuel supplies, and announced some across-the-board cuts in schedules.

On November 11, Interior Secretary Rogers C.B. Morton said that the chances of gasoline rationing were better than 50-50. But President Nixon indicated his opposition to rationing on November 17. He did, however, have standby rationing procedures developed, featuring coupons and other accouterments of wartime policies.

The Nixon administration announced a series of conservation measures, some of implemented by executive order and others by congressional legislation:

Industries and utilities using coal would be prevented from converting from coal to oil.

Jet fuel allocations would be reduced, forcing an estimated 10 percent cutback in air travel. Subsequently, jet fuel supplies were curtailed by an additional 15 percent.

Allocations of heating oil for homes, industries and schools would be reduced 15 percent from the 1972 usage levels.

Everyone was asked to reduce thermostat settings by 6 degrees to an average of 68 degrees. Temperatures in federal offices were to be set at 65 to 68 degrees.

Speed limits would be reduced to 50 miles per hour to achieve greater efficiency, and the use of mass transportation and car pools was encouraged.

The country would return to daylight saving time on a year round basis.

Herbert Stein, Chairman of the President's Council of Economic Advisers, issued a warning on November 29 that the energy crisis could result in a 6 percent unemployment rate, sluggish real economic growth threatening a recession in 1974 (which occurred), and inflationary pressures (which also occurred, with a vengeance). Subsequent studies were to show that the economy suffered losses in the tens of billions of dollars from the energy dislocations.

Yet although the Arabs were successful in getting the attention of the American people, they did not succeed in getting the United States to change its foreign policy.

Many political initiatives followed (and are summarized below), but they had relatively little immediate impact on the nation's dependence on imported petroleum. While the embargo lasted six months until March 1974, the United States continued to consume more oil. Indeed, whereas in 1972 the United States imported 4.7 million barrels of oil per day, by 1979 Americans imported close to 9 million barrels per day.[17]

In late 1978, the Iranian revolution led to the loss of Iran's five million barrels per day of production. Thereby the world oil markets were transformed from a position of slight production surplus to one of substantial shortfall. There was a repeat of many of the circumstances of the 1973 embargo, including long gasoline lines and major political recriminations. The stage was set for *doing something*, something with a more dramatic impact, something that will lead us to the beginning of this book's synfuels saga. Let's see how the political responses within the United States unfolded.

POLITICAL RESPONSE IN THE UNITED STATES

Although the embargo was directed at the United States and a few other countries, the effects of the volatility of the oil markets were felt worldwide. Oil is a fungible commodity, and was distributed by market forces rather than by Arab political objectives. During this period most other developed nations—virtually all of which were more dependent on imports than the United States—dealt with the higher prices and the shortages with reasonable pragmatism and minimum disruption. The United States, on the other hand, experienced significant economic dislocations. Why was the U.S. experience different?

U.S. Energy Political Climate

The United States, which had enjoyed relatively cheaper energy costs for many years, seemed to operate in a more highly charged political arena than others. American attitudes were shaped in part by general ignorance of the workings of commodity markets that cause price fluctuations. For example, a 1978 Gallup poll revealed that more than half of all Americans did not know that the county imported any oil.[18] Moreover, the public was prone to conspiracy theories concerning oil companies, as seen in a 1979 Associated Press-NBC News poll, which found that 68 percent of the public thought that the oil shortage was a hoax.[19] Unfortunately, congressmen were too willing to indulge such paranoia. One of Winston Churchill's trenchant observations regarding Great Britain applies: "The multitudes remain plunged in ignorance of the simplest economic facts, and their leaders, seeking their votes, did not dare to undeceive them."[20]

Dysfunctional political manipulation was all too evident in 1973, even before the tumultuous effects of the embargo. Because of distortions caused by energy price controls, cumbersome regulations, and rapid growth of consumption, the United States was already experiencing tightness in many energy markets in the summer of 1973, with gasoline refinery capacity particularly short. Moreover, when Congress reconvened after the transformative events in the oil producing countries, prices of domestic "uncontrolled oil" as well as imported oil had risen sharply since the prior year. When oil companies reported higher profits in next quarters, it was easy for congressmen to focus public discontent on the "obscene" profits of the oil companies by holding hearings. Notable among these were those chaired by Senator Henry "Scoop" Jackson (a Democrat from Washington State), the chairman of the Senate Interior Committee, which had been studying energy matters for two years. In June 1973, he asked the Federal Trade Commission to make a study of charges that the oil companies conspired to create the shortage. A week later he announced that the Senate Permanent Investigations Subcommittee on Government Operations, which he also headed, would investigate the possibility that the gasoline shortage was the result of a premeditated plan. Yet, astonishingly he was on record that the energy crisis was real: a few months earlier he had called it "the most critical problem—domestic or international—facing the nation today."[21]

Similarly, Representative Les Aspin (a Democrat from Wisconsin) alleged "There is little doubt that the so-called gasoline shortage in the Midwest is just a big lousy gimmick foisted on consumers to bilk them for billions in increased gasoline prices."[22] And Senator James Abourezk (a Democrat from South Dakota) wrote to President Nixon: "Our energy 'crisis,' I believe, is deliberately contrived by the major oil companies to achieve a number of objectives they are seeking."[23]

Those remarks made during the summer of 1973 before the disruptions of the Yom Kippur War made it seem that Congress was also oblivious about the causes of the new disruptions. Senator Jackson, this time as Chairman of the Senate Permanent Subcommittee on Investigations, was again in the lead of this movement. He launched hearings in the following January to which executives of the nation's seven largest oil companies were summoned: Exxon, Texaco, Mobil, Standard of California, Shell, Standard of Indiana, and Gulf. They were lined up to take an oath and subjected to hostile questioning under television lights. (A similar charade was repeated in 2008.) The opening political remarks suggested that the crisis was contrived and the shortages exaggerated. The atmosphere was hostile and the witnesses were bombarded with detailed questions outside the areas in which they had been requested to prepare testimony. Senator Abraham Ribicoff (a Democrat from Connecticut) accused the companies of cheating and "reaping the whirlwind of 30 years of arrogance" in taxes and other matters. He also charged that they misled the American people and deliberately created a panic situation to freeze out small independent companies.[24]

When the factual studies came in, there was, of course, no substantiation for any of the wild charges. But by then nobody cared. The facts showed that the profits of the oil companies, in terms of rate of return on investment, lag and have always lagged behind the average for large industry in the United States over any period of time. Unfortunately large profits after sharp run-ups in oil prices misled the public and the media, who should (and probably do) know better, reinforce such political behavior. That was a typical pattern, one we have since seen in many other situations.

To dig below the surface, a basic knowledge of economics pays off.[25] When prices first rise, the companies are initially better off. But when oil prices fall, which they do from time to time, they aren't. Moreover, much of the apparently impressive profit rise is illusory in terms of actual financial benefit. For example, when prices increase, accounting rules make the companies increase the value of the oil in inventory, and this shows up as profit. But since they have to replace that oil with equally expensive oil, the firms are not financially better off; indeed, are worse off because they are taxed on the inventory profits. Ruth Sheldon Knowles cited results from a study by the First National City Bank for 97 petroleum companies. The study showed "that profit return on equity increased from 10.8 percent in 1972 to 15.6 percent in 1973, which was only slightly greater than the 14.8 percent average return of all manufacturing. However, over the previous ten years the petroleum group averaged 11.8 percent compared with 12.4 percent for all manufacturing."[26]

Other than higher prices caused by events abroad, much of the domestic dysfunction in energy markets resulted from misguided congressional action and not by any misbehavior of the oil companies. In particular, governmentally imposed price controls inevitably resulted in shortages, rationing, and collateral economic dysfunction. A vivid example of such congressional action involves the case of natural gas in the 1970s timeframe as articulated by Donald Rice, president of the Rand Corporation and chairman of the National Commission on Supplies and Shortages:

> The roots of the gas shortage go beyond the coldest winter in decades—namely to the refusal of Congress to permit deregulation of natural gas prices…. With the price of natural gas kept arbitrarily low, the effect has been simultaneously to inflate demand and squeeze supply—classic conditions for inducing a shortage in any commodity. The harsh winter brought a surge in demand that would have been troublesome in the best of market circumstances. Coupled with long-term government controls, it triggered a crisis that had long been predicted. The crisis prompted Congress to permit reallocation of existing supplies, through a form of temporary deregulation. Thus we have come full circle: Government acts to allocate the short supply of a commodity that it helped to make scarce in the first place. The question now is whether Congress will address the real solution, namely, the permanent deregulation of natural gas prices.[27]

As an aside, natural gas prices were finally deregulated during the Reagan Administration. Consumer advocates fought this action claiming that the industry would price gouge. Needless to say, prices fell rather than rose thereafter.

There are instances where the government should intervene in the market for the benefit of society as a whole, but it needs to be clear as to why. Throughout most of the seventies, however, the government often played a counterproductive role. It continued to regulate petroleum prices—thereby stimulating the consumption of petroleum, at the same time that it taxed the resources of the oil industry hampering new investment. As a result, despite the consideration of much legislation to strengthen the nation's energy security in the years following the embargo (3,000 bills were introduced the following January), as noted previously, the level of imports rose substantially between 1973 and 1979 to the point it constituted half of domestic consumption.[28]

Congressional Action

In this highly charged arena, Congress turned to taxing the profits of the oil companies: the Tax Reduction Act of March 1975, which despite its title, had provisions that wiped out the 22 percent depletion allowance tax incentive for all but the smallest operators and tightened foreign tax credits. This action increased the federal tax on the oil industry by about $2 billion in 1975, which had the immediate effect of reducing industry funding for exploring and drilling for new oil. Even though President Ford signed this act into law, he quickly introduced legislation to phase out price controls on crude oil over a two-year period to improve cash flow to industry. But Congress spent the rest of the year arguing how further to extend controls instead.[29]

Outside the area of price controls and taxes, Congress did take some actions directed at furthering conservation and improving energy supply during the next two years. In the area of conservation, it passed an Emergency Energy Act, which provided 55 miles per hour speed limits and year-round Daylight Savings Time. But more substantive measures such as mandatory conservation and switching utilities from oil to coal got caught up in the more heated debate concerning price controls.

On a more positive note, in November 1973, Congress ended the five-year court battle in which environmental forces were attempting to halt construction of the Alaska pipeline. As a consequence, about 1.2 million barrels of oil a day began to flow to the lower 48 states by 1977. It also passed an act that relaxed Clean Air Act standards to allow power plants to convert to coal providing that primary air standards were still met. In addition, Congress passed legislation trying to accelerate the leasing of offshore lands for oil exploration and production. But the stringent environmental requirements included in the act had the effect of slowing the development of new reserves.[30] Leasing of offshore lands has remained contentious to the present day with the net result that opposition by environmentalists has kept vast reserves of petroleum and natural gas off limits despite decades of evidence at home and abroad that there is little risk of significant environmental damage using modern technology.

With regard to government organization of energy matters, two major actions were taken. A Federal Energy Administration was created in mid-1974. And, in early 1975, the Energy Research and Development Administration (ERDA) was created by pulling together the energy research efforts of the entire federal establishment. At the same time, Congress increased appropriations for federal energy R&D by 64 percent over the prior year. In addition, the Energy Policy and Conservation Act of 1975 created the Strategic Petroleum Reserve that was ultimately to store between 500 million and one billion barrels of oil. This reserve was to give the nation more flexibility in meeting political situations that might involve an interruption of the flow of imported petroleum.

Nonetheless, when the Ninety-fourth Congress adjourned in October 1976, Congressman Jim Wright (D-Texas) stated in the Congressional Record, "Since the Arab oil embargo three years ago we have tried to do a few timid things to reduce consumption, but they have not been very successful. We have dabbled with oil and gas pricing. We have made more money available for long-range research, for things like solar energy that may help us 30 or 40 years from now. But as far as doing anything practical to increase the supply of energy and reduce our dependence upon foreign sources in the foreseeable future, we have done nothing."[31]

President Carter's Early Initiatives

Jimmy Carter announced at the beginning of his presidency that energy would be one of his high priority areas. He charged James Schlesinger with developing a national energy plan on an expedited basis—aiming for delivery in April 1977, just three months after Carter's inauguration. This effort coincided with one of the coldest winters in a century, during which curtailments of natural gas (produced by regulatory actions) caused layoffs of the order of 425,000 workers in 7,000 closed factories. Many went without heat.[32]

On April 19, in a national television address, President Carter stated "Our decision about energy will test the character of the American people and the ability of the President and the Congress to govern this nation. ... This difficult effort will be the 'moral equivalent of war'—except that we will be uniting our efforts to build and not destroy...." His specific goals for the year 1985 were to reduce the annual growth in energy demand to less than 2 percent; to reduce gasoline consumption by 10 percent below the then current level; to cut oil imports in half, from a potential level of 16 million barrels per day to 6 million barrels per day; to establish a strategic petroleum reserve of one billion barrels; to increase coal production by about two-thirds; to insulate 90 percent of American homes and all new buildings; and to use solar energy in more than 2.5 million homes.[33]

The specifics of the president's program involved substantial taxation to force conservation—up to $5 billion a year near term, and $75 billion annually in the mid-1980s. The government would then return much of the money in the form of tax credits and incentives to those who undertook conservation efforts.[34] The key proposals were:

Imposition of a gasoline tax of at least 5 cents per gallon each year that national consumption exceeded stated targets. A ceiling of 50 cents was proposed,

Immediate deregulation of oil and gas prices was opposed as "disastrous for our economy", but newly discovered oil prices would be allowed to rise over a three-year period to the 1977 world market price, with allowance for inflation,

A new crude oil tax at the well head would be created over three years to equal the difference between controlled prices and world prices,

The crude tax and the gasoline tax would be rebated to taxpayers in an equitable manner,

A "gas guzzler" tax would be imposed on cars with above standard gasoline consumption,

The price of newly discovered natural gas would be raised to $1.75 per thousand cubic feet. It would apply to gas sold under new intrastate contracts as well as to new gas,

Industry and utilities would be prohibited from burning petroleum and natural gas in new boilers,

There would be a review of the NRC licensing process for nuclear plants to insure its objectivity and efficiency,

Utility rates would no longer contain promotional rates that did not reflect the true cost of energy,

Residential owners would receive tax credits for insulation and solar installations, and

A comprehensive reporting system would be imposed on all oil and gas companies.[35]

This latter part of the legislation was highly controversial and went through an extended congressional debate. Not controversial was the proposal to create a new Department of Energy that pulled together all of the government's energy functions. On August 4, the President signed the legislation creating the new department and James Schlesinger was confirmed by the Senate as its first secretary.

Despite acrimonious debate, the House acceded to most of the president's proposals, with the exception of the gasoline tax, the weakening of measures on utility reform, and the gas guzzler tax. His proposals had a much more difficult time in the Senate. The President addressed the nation on at least two occasions to put pressure on the Senate—to little avail. The final National Energy Act of 1978 contained the following much weaker provisions:

Deregulation of newly discovered natural gas by 1985 and sizable price increases in the interim. The ceiling price of new gas would rise immediately from $1.50 a thousand cubic feet to $2.09,

More modest tax provisions including a tax on gas guzzlers beginning in 1980;

Tax credits for energy saving weatherization devices; and some tax breaks for geothermal energy, solar, and wind energy equipment,

Requirements for utilities to burn coal rather than oil and natural gas, and, where feasible, for industrial plants to do the same,

Authorization for the Department of Energy to set efficiency standards for 13 types of major appliances, and requirements for utilities to give residential customers information about energy-saving equipment and to allow utilities to provide financing for such equipment.[36]

While these were steps in the right direction, much of the political response was unenthusiastic. The United States Chamber of Commerce prepared a study showing that the president's proposals were a disguised tax increase, "the biggest since World War II."[37] A Harris public opinion poll got a response of 56 to 32 percent agreeing that "the trouble with the Carter program is that it puts all the emphasis on conservation and very little on how to get new sources of energy." More telling was a statement by the National Association for the Advancement of Colored People, which said

> While we endorse the Plan's objectives of eliminating energy waste and improving utilization efficiency, we cannot accept the notion that our people are best served by a policy based on the inevitability of energy shortage and the need for government to allocate an ever-diminishing supply among competing interests. We think there must be a more vigorous approach to supply expansion and to the development of new supply technologies so that energy itself will not become a long term constraint, but instead can continue to expedite economic growth and development in the future.[38]

Ironically, no sooner did the National Energy Act get enacted than revolution came to Iran. At the end of October 1978, much of Iran's 6 million barrel per day production unexpectedly was interrupted because of strikes and turmoil in that country. On December 26, oil production ceased.[39] The General Accounting Office issued a report stating that the United States suffered a net reduction in supplies of 600,000 to 700,000 barrels per day during the first four months of 1979.[40] Saudi Arabia increased its production to make up some of the shortfall, but it was not enough to prevent the market from reacting with higher prices. By March, heating oil prices in the United States began soaring. Next, by June, OPEC responded to the increases seen in spot markets by setting new marker prices and surcharges that effectively doubled the price of crude to $23.50 a barrel from the $12.50 a barrel that obtained the prior October.[41] In parallel, gasoline prices jumped by 50 percent over a six month period. Worse, despite the increase in prices, gasoline lines were growing throughout the nation. California began rationing gasoline on an odd-even license-plate plan on May 9 to stem panic buying.[42] Again, independent truckers held a wildcat strike blockading truck stops on major highways in a dozen states to protest rising prices and scarce supplies. Carter signed a bill on November 5 giving the President authority to prepare a standby rationing plan and to put it into effect during a shortage.[43]

On another front, the President signed a proclamation requiring that air-conditioning in commercial, government and many other public buildings be maintained at temperatures no lower than 78 degrees Fahrenheit. Water in commercial buildings was to be no warmer than 105 degrees Fahrenheit[44] and in federal buildings all hot water to the washrooms was stopped. Yet oil company profits also inevitably grew during this period. Much of the profits were from overseas operations and much were inventory profits. Financial analysts pointed out still again that, over time, the profitability of the oil companies was below the average of United States industry. That did not stop President Carter from pursuing another tax increase—a "windfall" profits tax. He said that such a tax was the only thing that stood between the oil companies and a huge bonanza of unearned, unnecessary and unjustified profits.[45]

More Dramatic Action

Meanwhile the public's frustration increased with the lines and higher prices. The President was to make a speech on energy in early July, but this was cancelled. Two days later, a revealing memorandum from Stuart Eizenstadt, his chief domestic affairs advisor, was leaked to the press. He cited the angry public reaction to the lines and went on to say,

> I do not need to detail for you the political damage we are suffering from all of this. It is perhaps sufficient to say that nothing which has occurred in the Administration to date—not the Soviet agreement on the Middle East, not the Lance matter, not the Panama Canal treaties, not the defeat of several major domestic legislative proposals, not the sparring with Kennedy and not even double-digit inflation have added so much water to our ship. Nothing else has so frustrated, confused, angered the American people—or so targeted their distress at you personally, as opposed to your advisers, or Congress or outside interests. ...
>
> But I honestly believe we can change this to a time of opportunity. We have a better opportunity than ever before to assert leadership over an apparently insolvable problem, to shift the cause for inflation and energy problems to OPEC, to gain credibility with the American people, to offer hope for an eventual solution, to regain our political losses. We should seize this opportunity now and with all our skill... not one day or so of energy events can be allowed to pass without repeated follow-on events when you return from Camp David. Every day you need to be dealing with—and publicly be seen as dealing with—the major energy problems now facing us. Unless the attention to energy is almost total during the two-three weeks after your return, we will not turn the course of events around, and certainly we will not convince the American people that we have a firmer grasp on the problem than they now perceive.[46]

Eizenstadt went on to say that his staff would have concrete proposals for the president that would include the promotion of synthetic fuels and the creation of an Energy Mobilization Board. The proposal to be sent to Congress would call for the creation of an Energy Security Corporation designed to reduce imports by the year 1990 by 2.5 million barrels a day through the production of alternative fuels. This effort was to be funded from revenues received from the windfall profits tax.

The Energy Security Corporation eventually became the United States Synthetic Fuels Corporation. Because the account of the legislative process leading to the passage of the Energy Security Act, which created the Corporation, is so central to this book, the description of those events will be left to the full attention of Chapter III. Chapter II will first review the energy resources available to the United States to show why alternative fuels seemed like the answer to the energy "crisis" prevailing at the time.

II

The Potential of Synthetic Fuels

Thus, with the experience of a second disruption to petroleum supplies in a decade, the time seemed right politically to act decisively to improve the nation's energy supply security. More specifically, President Carter advocated for a substantial national effort to build from scratch an industry that would produce 2.5 million barrels a day of alternative fuels to displace imported petroleum. Why synthetic fuels?

Synthetic fuels are not as arcane as the name might suggest. They simply represent the conversion of one set of existing energy resources, largely solids, into forms more usable by the existing American energy infrastructure, such as liquids and gases. Such conversion seemed eminently sensible given that the United States' endowment of energy resources in solid forms vastly exceeds its resources in liquid and gas forms. We will look later more closely at the specifics. But, as an example, the United States has twice as much oil locked in the oil shale of three western states than there is oil in the Middle East and four times as much oil equivalent in coal deposits. Given these facts, analysts and policymakers asked why the United States should allow itself to be vulnerable to political blackmail from an unstable part of the world in order to supply its energy needs? And so attention turned to historical precedents for the production of synthetic fuels when energy availability became a national security concern.

PRECEDENTS

Notable precedents could be found in ambitious undertakings by Germany and South Africa only a few decades earlier. When each was confronted with the prospect of being cut off from world petroleum markets, they turned to synthetic fuels to meet essential needs for energy in liquid forms. For example, when the Nazis came to power in Germany in the early 1930s, they were already envisioning a major war effort for which petroleum would be essential. Although they could plan on exploiting both the modest Austrian oil fields (not brought into the Reich until 1938) and the much larger oil fields of Romania (to be occupied by the Germans in the upcoming war), these would by no means be sufficient for German military ambitions. Moreover, the fortunes of war might make imports from Romania problematic. Accordingly, German strategic planning turned to that country's large coal resources and German technological prowess (evident in the Bergius and Fischer-Tropsch gasification/ liquefaction processes) to meet the needs of its impending war effort.

These calculations turned out to be prescient. When Germany found itself cut off from world oil supplies due to the allied military embargo, it did receive most of its oil from occupied Romania (13 million barrels by 1941). Yet even though this accounted for half of Romania's total petroleum output, this production was inadequate for the German war effort. Moreover, the security of these reserves and production facilities was increasingly uncertain. Allied bombing and mining of the Danube reduced these levels of imports by half after 1943. Thus, as planners had anticipated, synthetic fuels production became central to fueling the German war machine. German synthetic fuels production rose from 10 million barrels in 1938 to 36 million barrels in 1943 (i.e., from 22 percent of total German consumption to more than 50 percent).[47]

Although South Africa did not have to contend with military necessity, its fears of a politically induced oil boycott prompted the government in Pretoria to develop a synthetic fuels capability in the 1950s. Their planners also proved to be prescient, although international pressure to restrict supplies did not occur until 1964, when Kuwait banned all petroleum exports to South Africa and OPEC followed suit in 1973.

In the intervening time, South Africa had brought a major synthetic fuel facility to completion, i.e., Sasol 1, in which two Fischer-Tropsch units operated for a design capacity of 10,000 barrels per day.[48] One used a fixed-bed catalyst like the one the Germans had employed during the war. This process, based on a proven technique, operated without difficulty. The second unit, an unproven fluid bed system, developed and installed by the American firm of M.W. Kellogg, required several years to debug before achieving design production levels. Both units are still in operation today.

The oil crises of 1974 and 1978 provided a substantial boost for Sasol when higher world prices made the cost of production more economically competitive. At the end of 1979, for example, Sasol's production cost was estimated to be about $30 per barrel at a time when world spot prices were about $10 per barrel higher. Sustained high prices led Sasol to embark on the construction of a second installation with a capacity of 40,000 barrels per day, Sasol 2, at Secunda in 1976.[49] By the time this installation came on-stream in 1980, work had already begun on another installation in Secunda, Sasol 3, which would produce an additional 40,000 barrels per day. When this third unit came on-stream in 1982, about fifty percent of South Africa's petroleum needs were met by synthetic fuels. Later on, the South African state-owned company built yet another facility in the 1990s. By 2008, South African synthetic fuels production amounted to 160,000 barrels per day.[50] These units now form the single largest and most profitable asset in Sasol's global portfolio.

Despite such precedents, American policymakers were unsure of the technology's promise. Other than the gasifiers built in South Africa using the German war-created Lurgi coal gasifiers, the technologies were unproven at commercial scale. They would be very expensive and they would take a long time to build. On the other hand, a period of sustained high oil prices would make the projects economical. Accordingly, the country had arrived at a point where it seemed worthwhile to devote national resources to confront the uncertainties and the risks of synthetic fuels development. The resources were clearly vast in amount and there were any number of technologies that had the potential to exploit the resources. Let's look at these more closely.

THE RESOURCE BASE

The U.S. resources having a potential to be converted into synthetic fuels—coal, oil shale, tar sands, heavy oil, and peat—could meet the nation's needs for fossil fuels for centuries.[51] Indeed, Table II-1 shows that coal and oil shale easily contain the equivalent of fifty times as much oil as the conventional petroleum resource (presented in billions of barrels of oil equivalent (BBOE)).[52]

These numbers were derived using the methodology recommended by the U.S. Bureau of Mines and the U.S. Geological Survey (USGS), which employs definitions of resources and reserves and their various subcategories in terms of a "McKelvey Diagram" (see Figure II-1).

According to this methodology:

A resource is a sufficiently concentrated form and amount of material to make economic extraction of a commodity from the concentrated material currently or potentially feasible. Several different types of resources are recognized, according to the degree of certainty of their existence: identified

PRODUCTION POTENTIAL ESTIMATES

Resource	Production Potential (BBOE)
Petroleum	27
Natural Gas	34
Coal	731
Peat	35
Oil Shale	626
Tar Sand Oil	8
Heavy Oil	33

Table II-1

Figure II-1 McKelvey Diagram

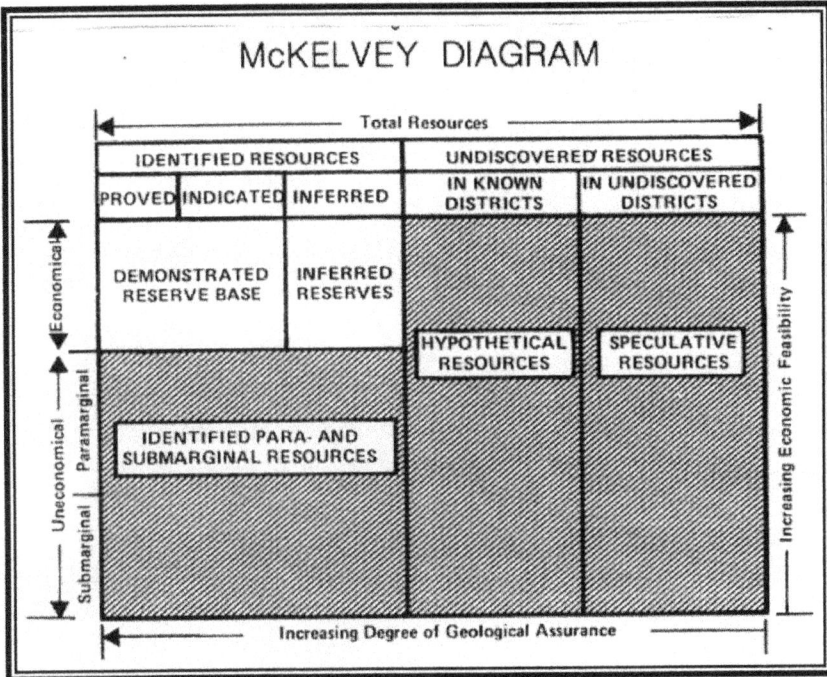

McKELVEY DIAGRAM

	IDENTIFIED RESOURCES			UNDISCOVERED RESOURCES	
	PROVED	INDICATED	INFERRED	IN KNOWN DISTRICTS	IN UNDISCOVERED DISTRICTS
	DEMONSTRATED RESERVE BASE		INFERRED RESERVES	HYPOTHETICAL RESOURCES	SPECULATIVE RESOURCES
	IDENTIFIED PARA- AND SUBMARGINAL RESOURCES				

Total Resources

(Economical) — (Uneconomical) / Paramarginal — Submarginal

Increasing Economic Feasibility

Increasing Degree of Geological Assurance

resources, inferred resources, and undiscovered resources. Undiscovered resources may be further subdivided into hypothetical resources and speculative resources. Hypothetical resources have a high degree of geologic uncertainty, but speculative resources have an even higher degree of geologic uncertainty. ...

The size of a resource always refers to the size of the resource in place, unless explicitly otherwise qualified. For example, the statement "The size of the identified and economic Eastern Interior coal resource is 213 billion tons" means that 213 billion tons of it are in place. The amount of ultimately recoverable resource would depend on mining or extraction method, losses during mining or extraction, legal restrictions, environmental considerations, subsidence considerations, etc. If all of the coal were located at depths that would require underground mining, the amount of coal recoverable could be as little as half of the resource in place.

Reserves are that part of the identified resource that could be economically extracted or produced at the time of the determination. Actual physical extraction/production facilities do not have to exist for the recoverable portion of the identified resource to qualify as a reserve. Detailed calculations, designs, plans etc. are not necessary to find a particular portion of a resource to be a reserve; it is sufficient if the resource has geologic or other characteristics similar to other resources currently being produced. The term "reserves" includes only the recoverable materials: that portion of the resources that may be brought to the surface. Expressions such as "recoverable reserves" or "extractable reserves" are redundant and should be avoided since they imply that there may be such a thing as "unrecoverable reserves".[53]

Another useful definition is the "production potential" of a resource, which considers the likely efficiency of extracting and converting the resource into synthetic fuel. For example, based on the knowledge of existing technologies, it appeared that about 25 percent of oil shale, once mined, would be lost in the conversion process of extracting the oil from the shale.

Employing this definition, the next sections examine each of the resource bases having synthetic fuel production potential.

Coal

Coal is the United States's most abundant energy resource. In 1975, the United States Geological Survey estimated that the size of the total domestic coal resource was a breathtaking 4 trillion tons. Most was categorized as undiscovered. Yet even a more realistic number of 675 billion tons is still large. A later analysis by the Synthetic Fuels Company, which allowed for loses due to inefficiencies of processing mined coal into synthetic fuel, estimated that a more accurate assessment of coal reserves for synthetic fuels purposes was 203 billion tons. Since one can derive about 3.6 barrels of oil equivalent (BOE)/ton, the production potential of the domestic coal resource base was estimated to be 731 billion BOE.

Figure II-2: Principal Coal Regions of the United States

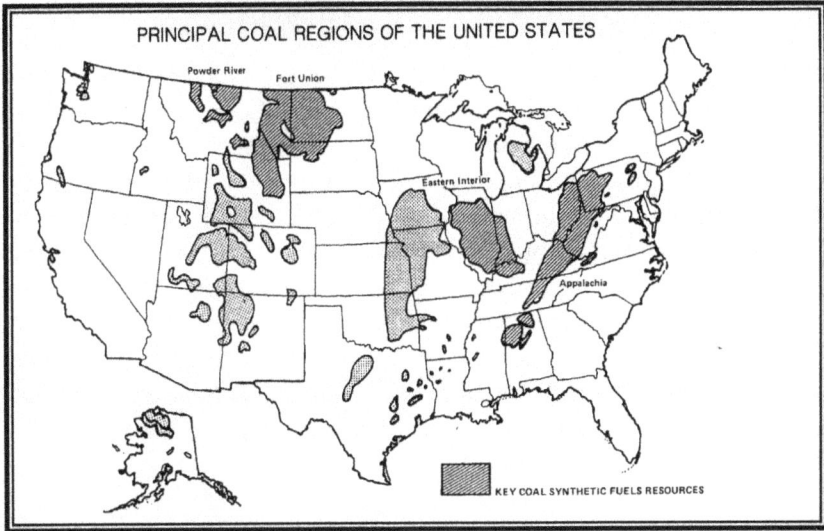

PRINCIPAL COAL REGIONS OF THE UNITED STATES

KEY COAL SYNTHETIC FUELS RESOURCES

These coal resources are widely spread across the United States as seen in Figure II-2.

Of these, there is considerable variability in the physical characteristics of the coal and its suitability for conversion into synthetic fuels. For example, coal varies considerably in its heating value and hardness, giving rise to the categories of anthracite, bituminous, sub-bituminous, and lignite.[54] Of these, anthracite constitutes less than one percent of the total, and can be neglected from further consideration in potential synthetic fuels production.

Peat

Although peat could be thought of as just a variation on coal (after all, peat is, in effect, a very young coal), its physical differences suggest different applications and handling requirements. Peat's major drawback is its high moisture content (90 percent or more). It takes as much energy to dry the peat as remains in it for practical use afterward, such that in most countries using peat, the sun is employed for drying. Also, compared to coal, peat generally has a higher oxygen content, higher reactivity, higher volatile matter content, lower sulfur content, and lower fixed carbon. Nonetheless, given peat's production potential of 35 billion BOE, comparable to all the domestic petroleum reserves, its potential role in synthetic fuels production could not be overlooked.

More than 80 percent of the nation's peat is located in just twelve states. Northern Tier states (Maine, Massachusetts, and Wisconsin) account for 31 percent, Southern Coastal states (Florida, Georgia, Alabama, South Carolina, North Carolina, and Louisiana) for another 12 percent in total; and Alaska for the remaining 41 percent. The Northern Tier resource would probably be the most attractive because that region has few indigenous energy alternatives, and because of significant drawbacks of the peat resource in other states. For instance, Alaska is too far from most energy markets to justify shipping the peat or its products, and the southern resource is too small and involves substantial coastal terrain likely to be environmentally controversial.

Oil Shale

Oil shale is a resource comparable in size to coal, with a production potential of 626 billion BOE. Oil shale is a rock, which when crushed and heated, releases kerogen, a raw shale oil that can be treated and processed much like natural petroleum. Figures II-3 and II-4 show the major shale resources that are found in the western and eastern regions of the United States, respectively.

There are important differences between the reserves in the west and the east. The western resource is both larger in quantity and richer in concentration. The eastern resource is more conveniently located and closer to water resources and labor markets.

More specifically, looking at Figure II-3, one sees that the western shale resource is found predominantly in three states: Colorado's Piceance Basin, Utah's Uinta Basin, and Wyoming's Green River Basin. The USGS has estimated that the western resource has about 2 trillion BOE in place. Much of this requires mining and some is very deep underground, and so the production potential may be only one-third of this amount.

The Mahogany Zone shale, which is found in the Piceance Basin, is the richest of the shale resources. Its shale contains as much

Figure II-3: Major Western U.S. Oil Shale Resources

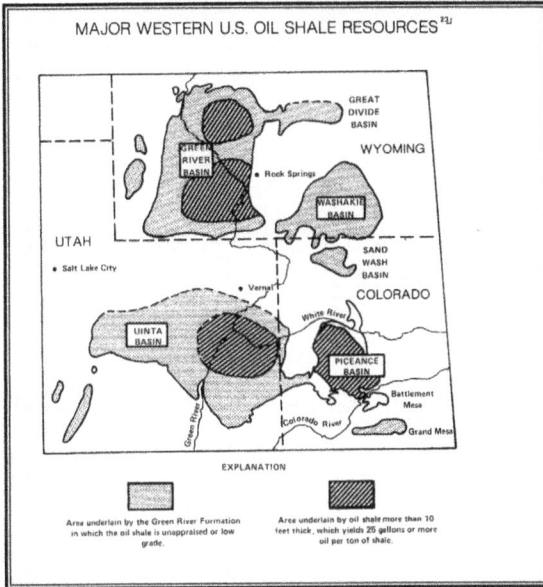

Source: D.C. Duncan and V. E. Swanson, "Organic-Rich Shales of the United States and World Land Areas." USGS Circular 523, 1965.

as 40 barrels of oil per ton. Utah's shale has a richness closer to 24 barrels per ton, whereas eastern shale has a richness of only half that amount. The Mahogany Zone is a large saucer-shaped deposit: its outer edges actually are visible on the sides of mountains, its intermediate areas are near the surface allowing strip mining, and the middle is deep underground (about 500 to 1,000 feet). Each would require very different mining techniques, having in turn different economic impacts. The oil from western oil shale is relatively paraffinic, allowing high yields of diesel and jet fuel from the refining process.

As seen in Figure II-4, the eastern shale resource is found over a widespread region including the states of Kentucky, Indiana, Ohio, and Michigan. Ease of mining suggests that the outcroppings of this resource are the portions most likely to be used in synthetic fuels production. While Fischer assay (a method for measuring the richness of oil content) shows eastern shale to be significantly leaner than the western shale, the carbon content of both are comparable. Thus, eastern shale could be competitive given the development of

Figure II-4: Eastern Oil Shale Resource

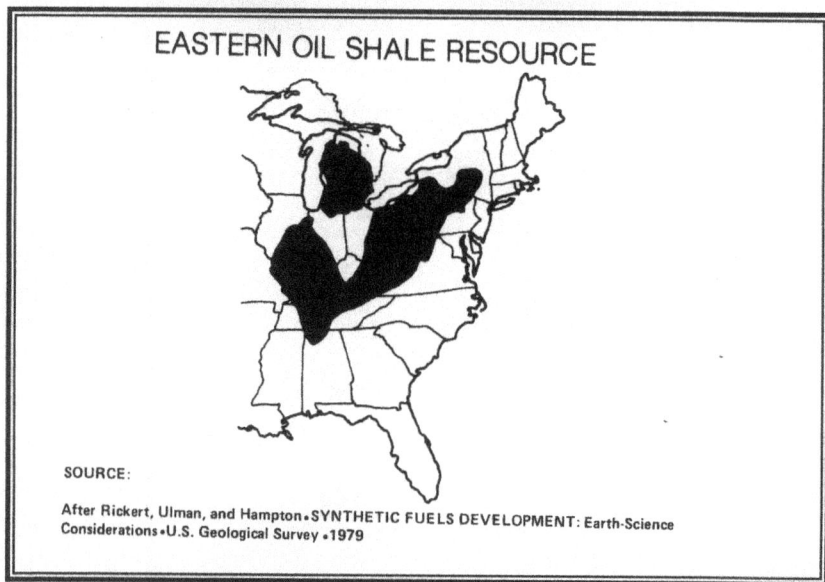

EASTERN OIL SHALE RESOURCE

SOURCE:

After Rickert, Ulman, and Hampton • SYNTHETIC FUELS DEVELOPMENT: Earth-Science Considerations • U.S. Geological Survey • 1979

technologies that would permit the carbon content of the eastern shale to be fully exploited.

Tar Sands and Heavy Oil

Tar sands and heavy oil resources are part of a continuum of oil resources that begins with light oil that flows easily from the ground matrix that holds it to progressively heavier and more viscous oils that will not flow on their own or through conventional oil pumping techniques. The dividing lines between one category of oil resource and another is a matter of scientific convention.

The definitions eventually employed by the Synthetic Fuels Corporation were:

Tar sand oil: any oil produced by mining, as well as oil produced by conventional means (wells) where the oil has a gas-free viscosity of greater than 10,000 centipoise at reservoir temperature. If viscosity data are unavailable, then any oil heavier than water (10 degrees API or less).

Heavy oil: any oil not a tar sand oil and having a density of 20 degrees API or less.

Considerable heavy oil was already in production in the United States using secondary and tertiary recovery methods such as steam flooding. Eventually, the synthetic fuels program would have to distinguish between heavy oils available through existing technology and those requiring new (i.e., synthetic fuel) technology development.

The domestic tar sand resource exceeds 22 billion barrels of oil. Of this amount, more than half is in Utah. Much of the remainder is found in California, Alabama, and Kentucky, where a relatively small portion is recoverable by surface mining techniques. As a result, the production potential was estimated to be only 8 million BOE. (It should be noted that Canada has enormous surface tar sand resources in Athabasca—i.e., 1.6 trillion barrels in place, which is why Canada already has a commercial tar sands industry.)

Domestic heavy oil resources have been estimated to be more than 100 billion barrels in place. More than half this resource is in California. Alaska has 25 billion barrels. Texas, Oklahoma, Wyoming, Arkansas, and Mississippi each are estimated to have more than one billion barrels. The amount of the heavy oil in place that will eventually be susceptible to production will, of course, depend on the technologies developed to recover them. The SFC's *Comprehensive Strategy Report* of June 1985 estimated in its appendix J that about a third of the resource would be recoverable, namely 33 billion barrels.

TECHNOLOGIES

Thus, even though this vast resource base is obviously enough to meet the United States's demand for fossil fuels for centuries, the energy contained in this resource base is in the "wrong" form (generally solids rather than liquids) and the technology was not available in proven, commercial-scale to provide the alternative forms of energy economically.

Considerable knowledge and experience regarding potential technologies were available. The South Africans had brought to a commercial scale (after much expense and extended technical difficulties), and the U.S. government had been supporting the development of new coal conversion technologies through the pilot scale for three decades since the end of World War II. Moreover, private firms had done much pilot work on oil shale technologies, such as Union Oil's rock pump, Tosco's ceramic ball technology, and Paraho's gravity feed technology. But none of these were commercially proven except for the Lurgi gasifier, and the Lurgi experience demonstrated that scaling up technology takes considerable time and expense.

Consequently, when Congress considered synthetic fuels legislation, many corporations said that they would be willing to build vastly more expensive commercial facilities if the government shared the costs and the risks.

The following sections discuss synthetic fuels technologies that were considered to be potentially economic for exploiting each of the key resource areas.

Coal

The primary interest in potential synthetic fuel technologies was the ones that could produce liquids to meet the needs of the transportation sector in lieu of petroleum. Nevertheless, given that the

Figure II-5: Coal Conversion Processes

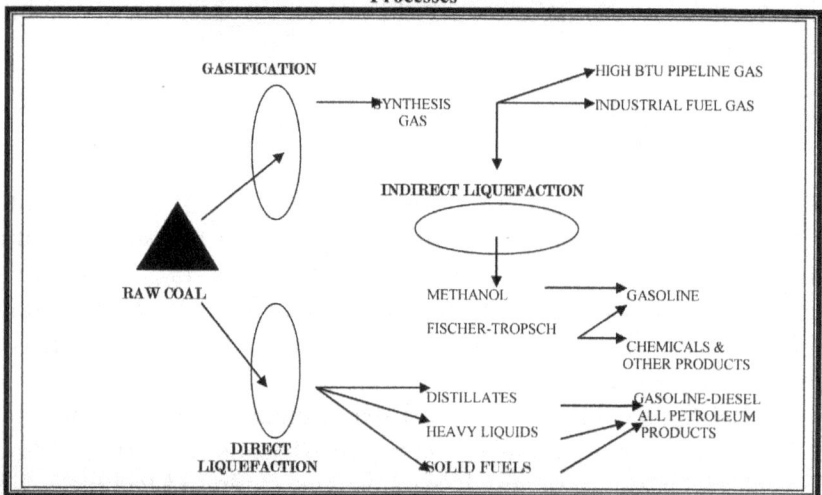

nation's gas resources were also seen as limited and given that gas can easily be converted into liquid fuels, it makes sense to look at technologies that can provide alternatives to both petroleum and natural gas.

Figure II-5 shows schematically the two fundamental approaches to convert coal to liquids, a direct process and an indirect process first producing gas. The gasification route first produces a gas equivalent to natural gas and then in a second step converts the gas to liquids, when that is the desired final product.[55] Accordingly, this route is termed "indirect" liquefaction. Direct liquefaction does not involve these two-steps.[56] Appendix A summarizes primary and secondary coal conversion technologies.

The chemical purpose of either approach is to add hydrogen atoms to the carbon atoms of coal, which can result in either liquids or gases. For example, in the gasification process hydrogen atoms from steam and oxygen atoms from air or pure oxygen interact with carbon atoms of coal to produce various gases, such as carbon monoxide (CO), hydrogen (H_2), and varying amounts of methane (CH_4). These gases can then go through secondary conversion processes to produce desired gas or liquid end products. Gasification technology, which entails ways to interact coal with steam and oxygen under conditions of elevated temperatures and sometimes elevated pressures, has found three basic approaches to be most promising:

Fixed-bed gasifier, in which coarse coal is processed by passing steam and oxygen upward through a bed of coal that moves slowly downward as it is consumed.

Entrained flow gasifiers, in which fine coal particles are carried through the reactor in a gas stream of higher gas velocities and finer particle sizes.

Fluid-bed gasifiers in which coal is kept in suspension in the gas with lower gas velocities and coarser particle sizes.

There are a number of proprietary approaches to these generic types, some of which will be covered in later chapters dealing with the specific projects considered by the Synthetic Fuels Corporation for financial assistance.

Research and development on direct liquefaction has pursued approaches of adding hydrogen atoms to carbon directly through processes involving high temperatures and chemical conversion by the use of catalysts. At the time that the synthetic fuels program was being considered, only a few proprietary approaches have been developed at pilot scale. For example, the Department of Energy and its predecessor agencies supported the Exxon Donor Solvent and the "H-Coal" processes in plants on the order of 200 tons per day.

Since peat can be gasified like coal, the gasification technologies listed above can also serve to produce alternative fuels from peat.

Oil Shale

Conceptually, technologies for producing synthetic fuels from oil shale are simple and many approaches have been considered at the laboratory and pilot plant scales. The various approaches involve crushing the shale, bringing the feed to at least 900 degrees Fahrenheit, and collecting the kerogen as it runs off. While the concept is simple, the technical execution, which ultimately determines the economics, is fraught with uncertainty. As a practical matter, technologies involving massive solids handling are not easily modeled. Scaling those technologies up from pilot to commercial scale is always difficult.

Potential shale technologies are distinguished from one another by how the shale is brought into contact with heat, the direction of shale flow, and the source of the heat:

Fixed-bed design, where at any given time the crushed shale fills the retort solidly, experiencing a temperature gradient from top to bottom. Various approaches attempt to optimize tradeoffs involving mechanical simplicity with efficiency of heating and extent of oil recovery, for example:

Crushed shale is fed into the top of the retort and the fixed bed moves downward.

Crushed shale is pushed up from below via a *rockpump*, such that as the shale rises, it encounters hot recycle gas.

Drawing on the experience of other solids handling industries, linear or circular grates would be employed to convey the shale through the heating process.

Other technological approaches attempt to maximize recovery by retorting all the fines in kiln-like retorts: for example, the Tosco rotating kiln process, Lurgi's Screw Mixer process, and Chevron's Staged Turbulent Bed process.

In an entirely different approach that minimizes solids handling and the need for retorts, Occidental has developed a Modified in Situ approach that is designed for deep mines in locations where economics preclude bringing all of the shale to the surface for retorting in one of the above types of retorts. A small fraction of the shale is mined to create a void, at which point explosives rubblize an associated volume that expands into the void. Air is blown through the rubble and the shale is ignited. The heat from the burning causes the kerogen to flow and it is collected at the bottom of the burning cavern and brought to the surface. This approach was designed for the center of the Mahogany Zone where rich deposits of shale are found at depths on the order of 1,500 feet.

Most of technology development for converting oil shale had been in the retort, but considerable pioneer effort was still required in the mining and the upgrading facilities as well (before the Parachute Creek Project went into operation with support by the SFC as discussed in later chapters). Although the industrialized world has extensive experience in underground mining, that experience is with the behavior of specific strata and minerals. In 1979, there was no such experience at full scale with shale. And shale mining was likely to dwarf other experience. Oil shale mines were expected to be 5 to 10 times larger than any existing coal mine. Shale mining would entail the digging of caverns by the room-and-pillar method that would have heights and widths of 40 to 60 feet.

Finally, appropriate upgrading technologies were needed to convert the raw shale oil to a usable product in refineries, a shale syncrude. Specifically, upgrading processes were needed to reduce contaminants, largely: nitrogen, oxygen, shale fines, as well as trace amounts of arsenic and iron. Also, some hydrogen must be added through catalytic processes. The refining industry has lengthy experience with catalytic processing, but experience was needed with the contaminants specific to shale. Once the shale oil is upgraded, however, a superior syncrude would be produced, one particularly suited for jet fuel and diesel fuel production.

Tar Sands and Heavy Oil

Because tar sands and heavy oil occur in physically similar forms in nature with the distinction between the two often arbitrary, it is more useful to discuss respective extraction technologies in terms of surface processing and in situ categories:

Surface processing involves the initial step of mining, which is followed by either retorting or solvent extraction processes. Retorting involves the same principles as shale retorting. The ore is heated in the range of 900 degrees F to drive off and crack the oil.

Solvent extraction, on the other hand, uses a solvent to dissolve the oil and separate it from the sand. The oil and the solvent are then separated and the solvent is recycled for further use. The Canadian tar sands industry is able to use hot water rather than solvent to separate the oil from the sand, but unfortunately, the deposits in the United States have different chemical bonds between the oil and the sand such that the Canadian experience is not applicable.

In contrast to mining and surface processing, in situ methods involve some externally applied, sub-surface stimulus to allow otherwise highly viscous tar or heavy oil to flow to the surface. Basically, viscosity can be reduced by heating the reservoir or by using additives. In addition, reservoir resistance to flow can be decreased by ground-fracturing and by increasing pressure.

Thermal methods include steam soak, steam drive, and fireflood (air or oxygen-driven). Other methods of reducing viscosity include CO_2 injection or chemical injection. Virtually all of the existing heavy oil production in California employs some form of steam soak. The choice of technology depends on the geology of the formation, its depth and the qualities of the oil being recovered.

Economic issues

By 1979, the United States had both a wide array of opportunities available and the political will to do something to alleviate any perceived energy vulnerability. The question was, what was the right policy? The United States were verging on a major commitment of national financial resources—the eventual program, as we shall see, was to be budgeted at $88 billion. If synthetic fuels were such a good thing, why should not their development be left to the private sector?

The arguments made in favor of a government-initiated synthetic fuels program were basically twofold. First, given market and technology risks, the private sector simply could not and would not undertake the development (at least, not soon enough to help Washington achieve energy independence). Second, there were substantial benefits to be gained by the nation from a synthetic fuels industry that were not reflected in the market dynamic, and that these benefits would significantly exceed the proposed costs. The underlying analysis for such arguments is presented at greater length in Chapter 10. Early versions of arguments along the following lines were considered when crafting synthetic fuels legislation.

Any first-generation plant built would not be economical, even at the seemingly high petroleum prices of the 1970s, because it is in the nature of pioneer plants that they cost more than original estimates, tend to be late in starting up, to have unexpected expensive problems to resolve, and to be years in achieving full capacity throughput. This makes the cost of output prohibitive to achieve a reasonable rate of return on investment. Given the necessary risk/ reward balance, investors would have to see very high rates of return to justify these major investments. Oil prices have until recently never reached the necessary levels to justify private risk-taking. The resulting high cost of output exceeds likely market prices and precludes a reasonable rate of return on investment.

But third-generation plants built fifteen years or so later might well be economical. This was because plants once in operation tend to improve their efficiencies strongly over time. The South African facilities were still experiencing increasing improvements in throughput, fifteen years after startup. One learns by doing. A plant design based on such experience would have much less risk and much improved economics than the pioneer facility. The financial argument for the government to bear some of the risk and startup costs of a synthetic fuels industry is that society as a whole would benefit from the learning experience. That is, having proven technologies with years of experience would yield the potential of building new generation plants in a much shorter period of time should the economics and the geopolitical situation justify it. In addition, the United States would be freer from potential political blackmail by oil-producing states when synthetic fuels technologies were perfected because they would provide a de facto ceiling on how high world oil prices could go.

Nevertheless, such arguments by themselves do not dictate the kind and scope of program to undertake. Should the country focus on proving a few key commercial-scale technologies, or should it aim for larger scale production? How should the government nurture such new technologies with minimal distortion to the private sector? Given the stakes involved and a sense of urgency, it would be all too easy to spend too freely and to allow political pressure rather than economics dictate the shape of a new industry.

These issues lay at the heart of the congressional deliberations over President Carter's proposed initiatives. The congressional action leading to the passage of the Energy Security Act is the subject of the next chapter.

III

Passing the Energy Security Act

President Carter's proposal to create a major synthetic fuels program to enhance the nation's energy security was not the first time the federal government had turned its eye to synthetic fuels. Indeed, earlier in the 1970s, Vice President Nelson Rockefeller proposed a vast $100 billion program to improve the nation's energy supply, and President Ford supported a $5 billion synthetic fuels program in 1976. But the former did not even pass committee consideration by Congress, and while the latter was strongly supported in the Senate, it failed by one vote in the House of Representatives. What was unique in 1979 was both that the president was willing to spend his political capital to reach this goal and the widespread feeling that the time was right.

This chapter shows how these conditions led to an unprecedented national endeavor by summarizing the president's proposal, outlining the legislative steps leading to the passage of the Act, and indicating the nature of the issues that confronted Congress along the way.

PRESIDENT CARTER'S PROPOSAL

On July 15, 1979, President Carter made a televised address to the American people in which he stated:

> Ten days ago I had plans to speak to you again about a very important subject—energy. For the fifth time I would have described the urgency of the problem and laid out a series of legislative recommendations to the Congress, but as I was

preparing to speak I began to ask myself the same question that I now know has been troubling many of you: Why have we not been able to get together as a nation to resolve our serious energy [problem]? ...

In little more than two decades we've gone from a position of energy independence to one in which almost half of the oil we use comes from foreign countries at prices that are going through the roof. Our excessive dependence on OPEC has already taken a tremendous toll on our economy and our people. This is the direct cause of the long lines that have made millions of you spend aggravating hours waiting for gasoline. It's a cause of the increased inflation and unemployment that we now face.

This intolerable dependence on foreign oil threatens our economic independence and the very security of our nation.

The energy crisis is real. It is worldwide. It is a clear and present danger to our nation. These are the facts and we simply have to face them. What I have to say to you now about energy is simple and vitally important.

He went on to lay out a six-point program along the following lines:

There would be a clear national goal established that the U.S. "will never use more foreign oil than it did in 1977."

Presidential authority would be used to limit imports to less than 8.5 million barrels of oil a day in 1979 and 1980. That level would meet the ceiling that the United States had just pledged at the Tokyo economic summit.

The Energy Security Corporation would be created to produce enough alternative or synthetic fuels by 1990 to replace 2.5 million barrels a day of imported oil. In addition, there would be the creation of a "solar bank" having the goal of capturing enough solar energy to provide 20 percent of the nation's energy needs by the year 2000.

The nation's utilities would reduce by 50 percent their consumption of oil by the year 1990 by shifting to other sources of power, primarily coal.[57]

The legislative process

The president's dramatic move energized a hitherto low-key legislative initiative for synthetic fuels. Since the upheaval in oil markets with the beginning of the Iranian Revolution some six months earlier, congressional committees had been exploring new energy initiatives.

Amendment to the Defense Production Act

For example, on June 26, 1979, the House passed the Senate initiative S. 932, which was designed to modify the Defense Production Act permitting assistance for the construction of synthetic fuels production facilities.

The Defense Production Act of 1950, passed during the Korean War, was designed to provide the executive branch with the necessary authority to ensure that defense needs received priority from American industry. Specifically, it was "An Act to establish a system of priorities and allocations of materials and facilities, authorize the requisitioning thereof, provide financial assistance for expansion of productive capacity and supply, provide for price and wage stabilization, provide for the settlement of labor disputes, strengthen controls over credit, and by these measures facilitate the production of goods and services necessary for the national security, and for other purposes."[58]

The amendment to the DPA sought to establish that secure energy supplies were essential to national defense needs. Moreover, it authorized the president to "contract for purchases of, or commitments to purchase, synthetic fuels for Government use for defense needs." Importantly, the amendment allowed the government to pay above market prices for the products, with two billion dollars authorized for this purpose. As voted out of committee, the bill set a production goal of 500,000 barrels per day by the year 1985 (later amended to 2 million bpd).

The amendment defined synthetic fuels as any solid, liquid, or gas that could be used as a substitute for petroleum or natural gas through a chemical or physical transformation of domestic sources of coal (including peat and lignite), shale, tar sands, and certain heavy oils. This definition of synthetic fuels would not change much through the various pieces of legislation considered up to the final passage of the Energy Security Act.

The associated bill was initiated by the Economic Stabilization Subcommittee of the House Banking, Finance and Urban Affairs Committee. It was known as the "Moorhead Bill," named after Congressman William Moorhead of Pennsylvania, who chaired the subcommittee. The legislation already had the support of Texas Democrat and House Majority Leader Jim Wright, who had supported past synthetic fuels initiatives. Indeed, Jim Wright had visited the president at Camp David prior to the speech and had emphasized the importance of synthetic fuels to the nation.[59] In part, it was the strong interest already shown in Congress that led the White House to announce its program.

S.932 as a Vehicle for the President's Proposal

Prior to the president's new initiative, the Senate had also considered promoting synthetic fuels. Senator Jackson, Chairman of the Senate Energy and Natural Resources Committee had introduced an Omnibus Energy Bill, which had some sections dealing with synthetic fuels.

Based on the work of an energy multi-agency task force, however, the White House concluded that the synthetic fuels initiative should go beyond any of the bills pending in Congress. Specifically, the task force had identified some 46 laws that would prevent the Energy Department or the Defense Department from moving quickly with a large program. This led to a desire for a quasi-independent financing corporation, using federal capital and the types of authorities contained in the Moorhead Bill.[60] But the White House decided not to draft a formal legislative proposal to Congress. Instead, it prepared informal specifications for the Senate Energy Committee's use, hopefully to be integrated into the pending Omnibus Bill.

At this point, the lead in legislative development passed to the Senate. The Senate Energy and Natural Resources Committee adopted most of the specifications with some modifications, two of which were especially important. First, many senators viewed the president's $88 billion program as too ambitious. Accordingly, there was a compromise approach involving a two-phase program: $20 billion to achieve 500,000 barrels a day, to be followed by a remaining $68 billion program subject to congressional approval of a comprehensive energy strategy.

The second modification was to increase the bipartisan appeal of the legislation by including an array of other energy initiatives in the Bill: renewables, alcohol fuels, solar, geothermal, and other alternate energy resources. So, for example, Senator Radley and others supported the Bill because of the alcohol fuels provisions. Senators Tsongas, Durkin, and Proxmire became supporters because of the solar energy program.[61]

In parallel with the above efforts on S.932, the Senate Banking Committee was taking a different tack, by drafting a Bill along the lines of the House's Moorhead Bill: providing a limited demonstration program based on narrow powers of the Defense Production Act, and appropriating $3 billion for loan guarantee authority. This amount was larger than it seemed however, in that it permitted three dollars of guarantee to be made for every dollar appropriated. The theory was that most loans would not default so that the appropriations did not have to cover the full amount of guarantees. Moreover, as loans were paid, they could be rolled over. Nonetheless, it represented a significantly curtailed program vis-à-vis the proposals favored by the president and the Senate Energy Committee.

The full Senate debated the two approaches at length from November 5 to 8, 1979. Finally, the Energy Committee version of S. 932 was adopted on a vote of 57-37. Many of the arguments over the two alternatives bore on critical elements of the Corporation and how the country should proceed with synthetic fuels (to be discussed in a subsequent section).

The Conference Committee and Final Enactment

Because the House and the Senate versions of legislation were not identical, the bills went to conference committee to resolve their differences. A total of 55 conferees were appointed, 32 from the Senate and 23 from the House. Congressman Moorhead chaired the

Committee, while Senator Johnston chaired the Senate conferees. The committee was the largest conference in congressional history. But Wright continued to play a key role because Richard Olson, his aide, became the committee's chief of staff.[62]

Given the scope of the proposed legislation, which eventually reached 350 pages in length, the conference lasted a grueling seven months, during which many changes were introduced. Some were superficial: for instance, the name of the quasi-federal institution was changed from the Energy Security Corporation to the United States Synthetic Fuels Corporation. Because the House version allowed the DPA authority program and the Corporation's program to run concurrently whereas the Senate just wanted the Corporation, a compromise was forged to terminate the DPA effort when the Corporation became fully operational (which not expected to take place until about a year after the passage of the legislation). In addition, the Corporation was not to have any authorities for biomass programs, which were moved in the legislation to the Departments of Energy and Agriculture.[63] Finally, a number of provisions were added to the legislation dealing with congressional oversight of the Corporation's activities to address the concerns of those congressmen who felt that the Corporation would be too free of scrutiny from either the executive or the legislative branches.

In the end, the revised bill racked up impressive majorities in its final passage. The Senate debate of the Conference Report took place on June 19, 1980: the vote was 78 to 12, with 10 not voting. The House debate took place on June 26, 1980: the vote was 317 to 93, with 2 answering present, and 21 not voting.

Nevertheless, many conservative members of Congress, particularly Republicans, remained unpersuaded that government had any appropriate role to play in developing the first generation of commercial synthetic fuel facilities. They believed that politicized investments rarely made economic sense; that this was something the government did not do well; and that the examples of the government in the synthetic rubber industry during the Second World War were irrelevant.

Prominent among these views was that of Texas Republican Ron Paul seen in the House debate on the Conference Report:

The promises of more abundant and cheaper energy for the taxpayer will never materialize, and when this is discovered the taxpayer is going to be very angry.

If the Congress and the bureaucrats can create consumer shortages when there are no actual shortages, how can we expect the same people to provide an abundance of energy from scratch? Seeking to turn these destroyers into producers is absurd. ...

Quite simply, we as Congressmen cannot know what is best and we cannot know what the market is saying—it is too complex. All attempts will fail and only serve to distort the information that true entrepreneurs rely on and are struggling with, to make some reasonable decisions. The big difference being it is their risk, not the taxpayers', thus prompting better judgment....It will be nothing more than a corporate CETA program, and is doomed to fail economically and become corrupt politically.

Because of reservations such as these as well as other issues, some of the compromises in the bill proved fragile over time and some key dissent lingered. To understand future events confronting the new corporation, some of the underlying issues that surfaced during the floor debates should be examined more closely.

Program Size and Scope

Eighty-eight billion dollars and some 30 to 40 facilities would be a massive national economic commitment. Did it make sense to build multiple copies of unproven technologies? Would the country not be better off to build a dozen plants to prove the technologies before replicating them? On the other hand, many argued that Washington had to show OPEC its determination, and that reducing U.S. reliance on imported oil justified the higher costs of a somewhat more inefficient approach.

Illinois Republican Congressman Tom Corcoran, who later came on the Board of the Corporation, supported doing only a limited number of pioneer plants, and voted against the Energy Security Act. Similarly, Congressman James Broyhill (R., N.C.) spoke during the House debate on the report:

Mr. Speaker, when this legislation passed the House last year, it was 14 pages long and authorized the expenditure of up to $3 billion in Federal money to assist private enterprise development of synthetic fuels. I supported that bill because I thought that it was a reasonable attempt to stimulate synthetic fuels production in this country. But now, 1 year after the passage of H.R. 3930, we are asked to swallow this legislation which now spans over 400 pages and which has a price tag which could be over $90 billion. To my budget-conscious friends I would like to point out that this bill poses the potential for busting every Federal budget over the next 10 years.

On the Senate side, Senator William Proxmire argued:

Mr. President, while I believe the conference committee on S. 932 has made considerable improvements in the bill which passed the Senate last November; I am still compelled to oppose this legislation. My problems are with title I, which would create not one, but two synthetic fuel programs.

When the Banking Committee considered the synthetic fuel legislation, we adopted an aggressive program which would have provided Federal assistance for up to 12 first-generation synthetic fuel projects. The Banking Committee felt that synfuels was an important energy option and that limited Federal financial assistance would help speed commercialization of this energy resource. But we did not feel that a crash program, seeking to obtain an attainable goal by the force-feeding of Federal dollars would represent a wise investment.

Separate Quasi-Federal Corporation

Others asked if the creation of a new quasi-federal entity justified or desirable. Some Congressmen felt it would be a boondoggle, while others that the lack of congressional oversight would be its undoing. Here are three views.

Senator William Armstrong (R-Colorado) in the Senate Conference debate said:

In fact, I think it very instructive, Mr. President, that over the months leading up to the passage of this legislation, the committees of the Senate and the other body heard much testimony from industry experts, economists, engineers, financiers, scholars, and consultants. We called these authorities in and, in effect, said to them, 'what will it take to get synthetic fuel production moving?' Interestingly, practically none of them suggested creation of the Synthetic Fuels Corporation or a similar massive bureaucracy. Instead they have recommended the following kinds of measures to cut red tape and provide incentives for the private sector ... ."

On this issue, Broyhill said in the House debate,

I am also deeply concerned about the workability of the U.S. Synthetic Fuels Corporation which title I of this legislation establishes. Under the conference agreement, this Corporation is not subject to the Administrative Procedures Act or the Corporation Control Act. Nor is it subject to many of the laws and regulations which govern the behavior of ordinary Government agencies or Government corporations. For that reason, the opportunity for congressional oversight over this Corporation's activities is likely to be minimal. As a result, we are practically inviting fraud, abuse, and arbitrary and capricious action on the part of this so-called Corporation.

Congressman John Dingell (D., Mich.) retorted by referring to improvements made in Conference over the earlier House and Senate versions:

More importantly, by requesting funds on an incremental basis, this change affords the Congress an opportunity to periodically review the operations of the Corporations, thereby assuring effective and full congressional oversight of the activities authorized by this Act ... .

A major failure in the Senate bill was the total absence of congressional control and oversight over the Corporation. The House conferees recognized this and insisted on creating an Inspector General responsible for monitoring the Corporation's activities and answerable directly to the Congress. This, perhaps more than any other legislative safeguard, will protect the taxpayer against fraud and inefficiency. I know our committee intends to use the Inspector General as a major part of our continuous legislative oversight of the Corporation.

Dingell's words are especially ironic in that the long-serving congressman was to employ Broyhill's view of the SFC during his numerous attacks on the corporation in subsequent years.

The Balance between Synfuels and Other Energy Initiatives

Finally, some of potential opposition to the creation of the Corporation was papered over in political compromises. Consequently, the Act received strong bipartisan support because it combined the synthetic fuels initiative with programs in other energy areas such as solar, biomass, and geothermal. While environmental groups tended to be hostile to synthetic fuels, they did not oppose the Act because of the other initiatives they held in high value.

Some of this was evident in remarks by Congressman Richard Ottinger (D., N.Y.) in the House Conference debate:

> Mr. Speaker, while I have grave reservations about some of the provisions of S.932, I nonetheless rise to support it … I am supporting the bill because it provides funding for conservation, solar and biomass efforts that I consider essential to resolving our energy crisis—and because I feel so very strongly that eliminating our heavy dependence on imported oil is the paramount need of the Nation, essential to our survival both economically and from national security and foreign policy standpoints… On the negative side, however, the bill provides far too much money, on an unreasonably accelerated basis and without adequate controls, for commercial production of synthetic fuels.

This balance was to prove precarious. One of the cardinal rules of high politics is that even though a bill passes by wide margins its opponents do not therefore become quiescent. Indeed, even in the face of demonstrable evidence to the contrary, they clung to their animus.

Thus, after lengthy debate and numerous reservations by key players, Congress had launched an innovative grand attempt at building a new industry. There was no comparable entity previously created in our form of government. The Corporation was to be free of much of the red tape that inhibited regular agencies and it was to be given unusual and very powerful financial tools to achieve the congressional mandate. How specifically was the legislation crafted to attain these ends? The next chapter examines the Act's key provisions.

IV

Provisions of the Energy Security Act

As outlined in the prior chapter, the passage of the Energy Security Act (Public Law 96-294) reflected a historic national political commitment. It created an innovative governmental entity and committed substantial funding on the order of the Marshall Plan (which had helped rebuild war torn Europe). What legislative provisions were contained in the Act to achieve its ambitions? Specifically, what were the tangible goals, the financial instruments, the administrative powers, and the funding authorized?

OBJECTIVES

The overarching objective of the Act, expressed at length in its introduction, was to expand the nation's energy production capability from alternative fuels so that the energy security of the nation could not be threatened. To that end, the Act set out goals for production levels and for diversity of energy resources and new technology.

Production Goals

The Act established production goals of 500,000 bpd by 1987 and 2 million bpd by 1992. These goals were driven more by the sense of urgency to achieve energy security than they were by considering the most economic way to build a new industry. While there was a political openness to the possibility that mistakes in the selection of projects could well be made in a rush to build the commercial production capability, there was relatively little consideration of whether the goals were physically and institutionally realizable.

To be sure, the drafters of the legislation realized that much was unknown and could not be known until more experience was had. Accordingly, the Act called for a two-phased program. The first $20 billion effort was to have been launched by 1984, and then a "Comprehensive Strategy" was to be submitted to Congress that drew on the experience of the first four years to provide a blueprint for the second and larger $68 billion phase. Despite uncertainties, Congress and the administration were determined to undertake a massive effort that could credibly improve national security.

The administration had already declared that achieving energy independence was "the moral equivalent of war." The principal consideration was undertaking an effort that would be commensurate with the perceived need and one that would receive the necessary political support while it rallied public attention.

One rule of thumb for crafting the legislation's production objectives seemed to be that if oil import levels could be brought down to around 2 million bpd, then the nation would be virtually energy secure because administrative measures such as rationing could allow the country to adapt to any energy interruption with an acceptable degree of economic dislocation. At this time, oil imports were averaging around 7.5 million bpd to meet national total consumption of about 18 million bpd. Extensive legislation had already been enacted to encourage conservation in the use of energy. These measures were estimated to result in a reduction in the domestic consumption of oil by about 4 million bpd, roughly half of the nation's imports. Thus, to bring total import levels down to 2 million bpd, additional production from alternative energy sources would be in the neighborhood of 2 million bpd.

A second concept was that economies of scale dictated that facilities be of a size around 50,000 bpd. A plant producing synthetic fuel from coal or shale was estimated to cost approximately $3 billion in 1980 dollars. Thus, a total capacity of 2 million bpd would require about forty plants at a total cost of about $120 billion. Assuming that the government provided three quarters of the cost and the private sector the remainder, the federal share would have to be about $90 billion. The private sector assured the congressional drafters of legislation that they were ready and able to carry out their share of a program. Notable among representatives of the private sector were American Natural Resources, Union Oil of California, and Tosco. Indeed, all of these companies proceeded with plants following enactment of the legislation.

In the abstract, the production goals were debatable. Wouldn't conservation efforts be more cost-effective? Weren't the world's economies so interdependent and oil so fungible a commodity that any effort on the part of the United States would be marginally insignificant? Possibly so. However, the goals represented as good an attempt to determine an amount that would make a difference as could practically be done. Nonetheless, such ambitious production goals were seriously flawed, as the actual experience of the SFC later showed.

Types of Resources/Facilities

It was easy enough to scope out the desired level of production and the estimated cost of achieving that production. However, the next obvious and more difficult questions would appear to be: what kinds of capacity were desired? Would it make any difference if they were all oil shale based? Should they all be 50,000 bpd facilities? Would it be politically acceptable if they were all located in Rocky Mountain states?

These questions were inherently unanswerable. Not enough was known about the technologies to analyze the economics of potential plants to give the answers any validity. Indeed, the purpose of the program, in part, was to scale up technologies, which had been developed through the pilot stage, to the size that promised economics compatible with commercial production—whatever that proved to be for a given technology and resource base. Ample evidence existed from other technology development (including earlier technology for synthetic fuels) that the economics could not be known short of designing, building and operating the actual facilities.

More relevant questions were along the lines of which technologies promised the greatest potential for meeting the nation's future energy needs. What kinds of plants were most likely to be economic, so that those needs could eventually be met in an optimum fashion? These questions could not be answered either, but they suggested how the Act should set specifications for program out-come and guide the Corporation to carry out Congress's mandate. The net result was that all technologies and resources sufficiently developed for private investors willing to place bets on them were to have a chance to compete. Specifically, the Act instructed the Corporation to make awards to those projects that:

Resulted in a diversity of technology and resource bases from among those made eligible by the Act. As mentioned earlier, the Act made the very largest domestic energy resources eligible, spe-cifically coal (including peat and lignite), oil shale, tar sands, and certain types of heavy oils. To the extent that a couple of differ-ent technologies could be successfully developed for each of these resources, the nation would never again have to fear shortages of oil and gas. Also, the approach improved the odds that the most economic forms of synthetic fuels would emerge from the group of resulting projects.

Were sponsored by "qualified concerns", which the Act defined as concerns that demonstrated the capability, directly, or through contracting, to undertake and complete the design, construction, and operation of a proposed synthetic fuel project. When this re-quirement was taken together with other requirements for spon-sors to incur risk through investing equity in a project, there was considerable built-in incentive for sponsors to propose the most economically appropriate projects in light of experience and com-mercial dictates of the market place.

Were most advantageous in enabling the program to achieve the national production goal established by the Act.

Would require the least commitment of financial assistance by the Corporation and the lowest unit production cost within a given technological process, taking into account the amount and value of the anticipated synthetic fuel products.

And whose proposals were most responsive to competitive so-licitations.

These requirements were designed to elicit participation by firms having confidence in their technologies, willing to share fi-nancial risks, and having a sound management track record.

FINANCIAL INSTRUMENTS

Having set the goals to be achieved, the Act's greater challenge was to provide commensurate financial means that established an appropriate balance of risk sharing between the government and private companies while offering sufficient incentive for companies to undertake first of a kind plants. While the government could have built government owned and managed facilities (as they did at the beginning of the nuclear industry), a higher priority was placed on leaving the initiative in the private sector, given that it had the expertise and talents necessary to insure that the plants were appropriately designed and managed.

Specifically, the applications of the technologies (e.g., synthetic natural gas, low or medium Btu gas, synthesis gas for industrial processes, methanol, gasoline, diesel fuel, and electricity) were a function of a complex and changing marketplace; the most appropriate feed stocks could be determined only in light of the needs of the technology, the end products, and the costs of alternative feed stocks (including transportation costs); and the location of the facilities were best determined considering available infrastructure, local construction costs, the distance to major markets, and feedstock availability. Because the unknowns were so substantial, indepth expertise and the corresponding intuition for addressing the commercial risks were needed. Moreover, the private sector, being more comfortable with dealing with the associated risks, was more likely not to over design the plants and was more likely to design commercially economic facilities—providing they were given the incentive to do so.

But how could the government's interests be protected when placing billions of taxpayer dollars of in the hands of private groups? The question was critical. Although private firms may have the requisite expertise, they also needed the right incentives to guarantee that they would serve the public interest. Therefore, the act's objective was to authorize financial assistance in the forms that would reconcile the government's needs and objectives with the capabilities and motivating factors of the private sector. It did so by allowing the government to assume market risks such as price fluctuations and availability of capital, while leaving technology and management risks in the hands of private sponsors.

Assuming that a corporation wished to build a facility whose economic prospects were at the time marginal, but which promised in a number of years to be economic, which aspects of the private capital market precluded the corporation from getting financing? These were essentially threefold:

The magnitude of the financing, which dwarfed the borrowing capacity of all but a few of the nation's corporations. If a facility were to have the prospect of being close to economic, it would have to be sufficiently large to capture full economies-of-scale. For most coal and oil shale technologies those facilities would cost in the neighborhood of three to five billion dollars. No corporation would wager the major part of its net worth on the success of a pioneer plant. Nor would any prudent lender provide non-recourse funding.

Completion risk—that is, the risk that the technology, when scaled up from pilot plant scale, might not work, or work at such low capacities that the project could not recover its capital much less provide a rate of return.

Market risk—that is, the risk that market prices might not rise as soon as the project anticipated and, thus, that the project might not for a period of time have the revenue stream required for it to remain viable.

Rather than providing funds directly to the private sector to build a synthetic fuels facility (such as by a grant or as shared equity), the Act authorized the use of loan guarantees and price guarantees to enable motivated corporations to overcome the above obstacles. Such financial mechanisms had been successfully employed in other government programs, but at much smaller scales. These instruments were to function as follows:

A *loan guarantee* places the full faith and credit of the federal government behind the loan so that the lender could be sure of the loan being repaid in the event of a project's default. Then, if the technology failed to perform, the lender would be protected. These guarantees were, however, accompanied by a key requirement to be sure that the proceeds from the loan were used wisely, and the guarantee did not function as an indirect grant to the sponsor's management. This requirement was that a loan guarantee could not be for an amount greater than 75 percent of the cost of a facility and that the sponsors had to have significant equity at risk in the project. For example, if a project were to cost four billion dollars, the government would guarantee up to three billion dollars and the sponsoring corporation(s) would have to raise the remaining one billion dollars in the form of private equity. (The Act also authorized as a less preferred form of assistance direct loans to be made by the government. Because a guarantee by the government is in virtually all instances sufficient incentive for private financial institutions to make loans, the direct loan feature, given its lesser priority, never came under serious consideration.)

A *price guarantee* establishes a product price level at which the project would presumably be economic and would provide the sponsor an adequate rate of return. Then, if the market price was in fact lower when the plant was completed, the government would pay the difference. If the market price were higher, then the government would not be obliged to make any payments. This mode of guarantee, taken alone, places substantial risk on a project's management. If the project fails to be built within the estimated cost, it is unlikely that the price guarantee level will provide the desired rate of return. If the project fails to function at all, the government is not obligated to make any payments. The Act also provides authority for the use of purchase commitments, which for all practical purposes, were a variant of the price guarantee incentive. The principal difference was that the government would contract to purchase the output of a facility and thereby ensure a market—conceivably assuming a market risk if the product were one not normally distributed in commercial channels. The commitment could be for a price exceeding market prices, and, thus, in practice would function very much like a price guarantee. Because all of the projects considered by the Corporation would have products with ready markets, this form of incentive was never seriously considered in negotiations.

Even though these forms of assistance could be potent, the universe of institutions that could actually make use of them was limited. On the positive side, this reality produced technically and financially sound projects. On the other hand, it left the impression that the private sector was relatively uninterested.

Although Congress could not be absolutely sure of the efficacy of such financial instruments, it was determined that the nation must obtain the desired synthetic fuels capabilities. As a consequence, the Act stipulated that the above forms of financial assistance were the preferred means of incentive, but indicated that if they proved to be ineffective in motivating the private sector to undertake the plants, then more potent forms of assistance could be employed — involving less risk-taking by the private sector and allowing more unilateral decision-making by the government. In order of priority, two additional forms of incentive were included in the authorities: joint ventures and Corporation Construction Projects.

Section 136 of the Act provided authority for the Corporation to engage in joint ventures with the private sector wherein the government would finance no more than 60 percent of the cost of a "module." According to this section, a module would be a facility of a size smaller than a full project, a facility, which, if successful, would demonstrate the technical and economic feasibility of the commercial production of synthetic fuels, and a facility that could eventually be expanded at the same site into a full-scale project. This provision was clearly intended to reinforce the Act's ability to achieve the goal of technical and resource diversity by permitting direct government investment.

But, even in this provision, the drafters of the legislation did not lose sight of the underlying rationale adopted for the design of the new industry — i.e., that the key design and management decisions be kept in the hands of those best positioned to judge the factors that would most likely result in commercial success. Accordingly, Section 136 (e) stated that "The Corporation participation in any joint venture pursuant to this section shall be limited to financial participation only and shall not include any direct role in the construction or operation of the module, other than that provided in subsection (f)." Subsection (f) merely clarified that this restriction did not otherwise prohibit the government from involvement in other partnership decisions involving the protection of the government's financial interest in the project.

Finally, if all the above forms of financial assistance failed to motivate the private sector to undertake projects that would accomplish the goals established by the act, then the Corporation was authorized under Subtitle E to own synthetic fuel projects, and the Corporation could contract for the construction and operation of such projects—"Corporation Construction Projects." The Act only allowed three such projects and indicated they could be built if appropriate projects could not otherwise be solicited to the extent necessary for the Corporation to achieve the Act's Production Strategy.

Among this array of financial instruments, the Act laid out clear preferences. Section 131 (b)(2) stated:

> The proposal selected for financial assistance pursuant to any solicitation shall be the proposal which, in the judgment of the Board of Directors, is the most advantageous in meeting the national synthetic fuel production goal established under section 125. Preference in such selection shall be given to—
>
> (A) the proposal which represents the least commitment of financial assistance by the Corporation and the lowest unit production cost within a given technological process, taking into account the amount and value of anticipated synthetic fuel products; and
>
> (B) in determining the relative commitment of the Corporation, in decreasing order of priority—
>
> (i) price guarantees, purchase agreements, or loan guarantees;
>
> (ii) loans; and
>
> (iii) joint ventures.

In addition, Section 131 (b) (3) (B)(i) directed the government to consider in awarding assistance "the potential cost per barrel or unit production of synthetic fuel". And, in using these financial instruments, the Act directed in Section 131 (a) that: "Whenever, in the judgment of the Board of Directors, it is practical and provident to do so, the Corporation shall award financial assistance on the basis of competitive bids."

The Act through the above array of financial assistance mechanisms provided powerful means for the program to achieve its goals considering most likely eventualities in the readiness of technologies, capabilities of sponsors, and the state of energy markets.

CREATION OF THE U.S. SYNTHETIC FUELS CORPORATION

The Act was as creative in its choice of institution to carry out the program as it was in its authorization of financial incentives. It authorized the creation of a new quasi-federal entity apart from the normal agencies of the executive branch—namely, the United States Synthetic Fuels Corporation. By this means, Congress took an unusual step to enhance the likelihood that the Act could be carried out unencumbered with the procedural constraints imposed on federal agencies. Moreover, the program was designed to be of limited duration and to require the skills of a staff not usually found in government.

Accordingly, the Act vested the powers of the Corporation in a Board of Directors, consisting of a chairman and six other directors appointed by the president with the advice and consent of the Senate. The chairman was to be a full-time employee of the Corporation with no other salaried position. He was to be appointed for seven years, and the other directors were initially to be appointed with staggered terms of from one to six years. Thereafter, the appointments were to be for a full seven-year term. A quorum of the Board was specified as a majority—i.e., four members. Any action by the Board required a majority vote of all the members; again, four members. This proviso would have severe consequences for the Corporation's ability to conduct business during two critical periods.

The Board did not, however, report to the president. Once appointed, the Board had sufficient authority to carry out the program. Moreover, the Energy Security Reserve provided the Corporation with the necessary "no-year" funding to cover its administrative expenses as well as to make assistance awards to projects without any further congressional action. In theory, as long as the Corporation operated within the law, it could carry out its mandate with an extraordinary degree of autonomy.

To be sure, the Corporation's actions were subject to oversight review by as many as seven congressional committees and subcommittees. In addition, the General Accounting Office was authorized to carry out studies of the Corporation as requested by Congress, and the attorney general could bring suit if it appeared that laws were being violated. The first two of these elements were employed vigorously and often by the Program's opponents. Nonetheless, although they could influence and harass, without further legislation, they could not force the Board to a course of action, or deter it from one.

Several other provisions of the Act contributed to the Corporation's independence of action:

Except as specifically provided in the Act, no Director, officer or employee was to be subject to any law of the United States relating to employment—i.e., were not subject to Civil Service regulations, thus greatly facilitating the staffing and administration of the Corporation. Further, Section 117(b) (2) authorized the Board to fix compensation of officers and other positions in excess of the Executive Schedule if such proposed actions were transmitted to the President and not disapproved by him within thirty days.

Section 175 stipulated that no Federal Law was to apply to the Corporation as if it were an agency or instrumentality of the United States, except as expressly provided in the Act. This provision led to relief from many procedural requirements faced by agencies in such matters as governed by the Administrative Procedures Act. Importantly, this freed the Corporation from all of the government's extensive procurement regulations.

The same section stated that no action of the Corporation excepting the direct construction and operation of a synthetic fuels facility shall be deemed a "major federal action significantly affecting the human environment" for purposes of ... the National Environmental Policy Act of 1969..." Accordingly, funded projects would not be required to conduct environmental impact statements in order for the project to proceed. All other environmental regulations at the federal, state, and local levels still had to be met. In addition, projects had to develop and implement Environmental Monitoring Plans after consulting with the EPA, the DOE, and state agencies. In effect, this provision reduced the upfront expense of a project, at a time where the government's commitment was not yet firm, and increased projects' costs over their lifetimes to maximize learning concerning the environmental effects of these new technologies.

Section 131(d) added protection for project sponsors by ensuring that, subject to the conditions of any contract for financial assistance, contracts shall be incontestable under law except as to fraud or material misrepresentation on the part of the holder. Congress was mindful of the substantial amounts of equity that the private sector would have to invest. Firms would not be likely to do so if the project's success could be hostage to the variety of suits our society has become increasingly adept in pursuing.

The above provisions all facilitated the Corporation's ability to carry out the objectives of the Energy Security Act. Congress, however, did not exempt projects from other provisions implementing public policy. Notable among these were requirements to have: projects receiving direct loans or loan guarantees certify that wages paid were not less than required under the provisions of the so-called Davis-Bacon Act; recipients of financial assistance provide for the fair participation by small and disadvantaged businesses in the synthetic fuel project; and projects receiving financial assistance to license their technologies to others under normal commercial terms.

Congress, aware of the possibilities for unwarranted action by the Board given its unusual autonomy, placed a number of key restrictions on its operations:

The financial disclosure provisions of the Ethics in Government Act of 1978 were made applicable to Directors, officers and many employees of the Corporation.

Employees of the Corporation were subject to substantially the same post-Corporation employment restrictions as would be other federal employees in similar circumstances—i.e., they would not be permitted to represent subsequent employers in front of the Corporation on any matter in which they had been materially involved.

All meetings of the Board of Directors to conduct official business of the Corporation were to be open to public observation unless the Board closed the meeting to avoid disclosing information that could adversely affect markets, could lead to speculation in securities, could impede the Corporation's ability to negotiate contracts for financial assistance, or could result in the disclosure of proprietary information.

The Corporation had to make available to the public, upon request, any information regarding its organization, procedures, requirements, and activities—excepting it could withhold information that could inhibit the Corporation's ability to negotiate contracts, or would result in the disclosure of proprietary or confidential material. This provision was very similar to, but not identical with, the provisions of the Freedom of Information Act, which governed the activities of federal agencies.

The Act established the position of Inspector General who reported directly to the Board to, inter alia, supervise auditing, investigative, and inspection activities relating to the promotion of economy and efficiency in the administration of, or the prevention or detection of fraud and abuse in, programs and operations of the Corporation. He was to give particular regard to the activities of the Comptroller General of the United States and to report expeditiously to the Attorney General of the United States whenever he had reasonable grounds to believe there had been a violation of federal criminal law.

Thus, Congress went to lengths to provide measures and to create a new institution to exercise those measures to a degree commensurate with the energy security problem they foresaw.

Number of Employees: Section 117(d) of the Act authorized the Corporation to employee up to 300 full-time professional employees to carry out the purposes of the Act. No limitation was placed on numbers of non-exempt support staff. At its greatest size (in 1984), the Corporation employed about 230 individuals, of whom approximately 60 percent were professionals. Thus, it never employed much more than one-half the number of staff permitted by Congress.

Salaries: The Act authorized the Board of Directors to fix the salaries of individual officer positions and categories of other employees taking into consideration the rates of compensation in effect under the Executive Schedule and the General Schedule prescribed under law for government employees. If, however, the Board found it necessary to fix any of those salaries at a higher level (presumably in order to hire qualified individuals) it could transmit its recommendations to the President, and if he did not disapprove within 30 days, such recommendations would go into effect. It was this provision, enacted by Congress itself that was used to paint the Corporation with the charge of profligate salaries despite it being used in only a few instances (see Chapter 12).

Funding

Congress was as generous in the funding of the program as it was expansive in the creation of authority. Public Law 96-126 enacted in November 1979 authorized $88 billion ($230 billion in 2010 dollars) to carry out the program and immediately appropriated $20 billion to the Energy Security Reserve. While some other alternative fuels efforts were authorized to be funded from the Reserve (e.g., alcohol fuels, waste energy, and the Solar Bank), the bulk of the funding was made available to the Corporation. Initially, the Corporation had something in excess of $16 billion available to make financial assistance awards and to cover its own administrative expenses.

With regard to the latter, the Energy Security Act authorized the Corporation to expend during any fiscal year $35 million for necessary and reasonable administrative expenses, and up to an additional $10 million for generic studies and for specific reviews of individual projects that had applied for financial assistance. These sums were to be adjusted each fiscal year by an amount equal to the percentage increase in the Gross National Product implicit price deflator for the year then ended.

These sums were ample for carrying out Phase I of the program. (Later, the Corporation determined in its Comprehensive Strategy Report that the $88 billion, which was not escalated with inflation, would not have been sufficient to meet the full production goals of the Act.)

In short, the program had much going for it—generous obligational authority, an independent institution dedicated to a national mission, and having political support from large majorities in both houses of Congress. How would these factors be used?

V

Getting Underway

The Energy Security Act provided unprecedented authority and funding to the new organization. At the same time, its congressional authors conveyed a sense of urgency to get the ambitious program underway, embracing a strong desire to avoid the cautiousness and approach of the normal government bureaucracy. Even though an election, a change of administration, and shifting energy markets introduced uncertainties to this picture, the Corporation got off to a fast start with regard to leadership, staffing, and identifying likely projects.

THE BOARDS OF DIRECTORS

Under the provisions of the Energy Security Act, a full Board consisted of seven members, one of whom was to be the chairman, a full-time chief executive officer of the Corporation. Over its relatively brief lifetime, three different Boards governed the Corporation.

President Carter nominated the Corporation's first Board of Directors in October 1980. Unfortunately, because of the administration's inability to reach a compromise with influential members of Congress about naming specific members (only four of whom could be from the same party), the Senate did not confirm Carter's nominations before adjourning for the 1980 elections. As a result, President Carter made recess appointments for all seven of the nominees, a constitutional maneuver that allowed them to serve until the end of the next session of Congress, a little over a year away.

President Carter named John Sawhill as Chairman. Sawhill was at the time the Deputy Secretary of Energy who had previously held positions as President of New York University, Administrator of the Federal Energy Administration, and as an officer of Commercial Credit Corporation. Those credentials in energy, government, and finance would be crucial to the success of the Corporation. Other Board members were distinguished as well, namely, John DeButts, retired Chairman of AT&T, and Lane Kirkland, head of the AFL/CIO. In addition, the first Board consisted of Cecil Andrus, Catherine Cleary, and Frank Savage.

This Board expeditiously approved an organizational structure, confirmed all the officers of the new organization, and issued the SFC's Initial Solicitation for projects to be considered for assistance in their first three months. In short order, staff, amounting to approximately a third of the Corporation's eventual maximum size, was hired.

Nonetheless, at the end of January 1981, ten days after President Reagan's Inauguration, he accepted the Board's resignation. The fact that none of them had ever tendered their resignations seemed to make no difference in the outcome. In fact, under the terms of the recess appointments, the Board members could have remained in office for another year regardless of the president's desires. After all, they were technically not members of the executive branch. In the event, the members obviously felt that with only recess appointments, no promise of any continuity, and some promise of enmity from the new administration, there was little point in not acceding to the White House's wishes. In departing, Sawhill designated Jack MacAtee, the SFC's General Counsel, as acting chairman until the nomination and confirmation of a new chairman. This interim period lasted until May 14, 1981, when Edward E. Noble was confirmed by the Senate as the new Chairman of the Board. The remaining six Board members were confirmed by October of that year.

The new Board's members were less distinguished than the original, but all had strong experience in the energy industry, either in production or finance. An additional important qualification was that Noble had the confidence of the White House and the new president, who was uneasy about launching a program twice the size of the Marshall Plan. He had been a fundraiser for the President's campaign and a major supporter of the Heritage Foundation, a conservative "think tank" just coming into its own. It was evident that the Administration needed assurance that the Board would undertake nothing imprudently extravagant.

While these aspects of Noble's background tended to receive the most notice, there were other areas of experience that suited him for the new position. Much of his wealth came from Noble Affiliates, a substantial independent oil and gas development firm. In addition, he had undertaken the development of a number of major commercial projects, involving hotels and shopping centers, with some success. He had an in-depth knowledge of the energy business from the ground up, and had been through the nitty-gritty of many commercial negotiations. While the Washington political community disparaged him for his relative lack of articulateness, he was shrewd where it counted.

The other six Board members came from a similar mold. They consisted of wealthy entrepreneurs, a banker to the oil and gas industry, and a commercial developer—four Republicans, two independents, and only one Democrat. They were:

Victor Schroeder, who was a commercial developer from Atlanta, Georgia. He had participated in a number of successful projects with Noble and had Noble's full confidence. Indeed, Noble arranged for the Board to appoint Schroeder as President, and chief operating officer of the Corporation. Noble had no desire to oversee the day-to-day detail of an operating institution and intended to look to Schroeder to fill this role.

Victor Thompson was chairman and chief executive officer of the Utica National Bank & Trust and Utica Bankshares Corporation of Tulsa, Oklahoma, with close ties to the energy sector. He had worked closely with Ed Noble in the past, and as long as he remained on the Board, he was regarded as the chairman's man (along with Schroeder). For a brief period, after Vic Schroeder resigned as president, he was elected president of the Corporation.

Robert Monks was an entrepreneur out of Maine, who had come to Washington with Vice President George H.W. Bush. He was past chairman of the board of the Boston Company, Inc., and former president of the Maine Wood Fuel Company, Ram and Company, C.H. Sprague and Son Company.

Howard Wilkins was another entrepreneur who had founded the Maverick Company, and was chairman of the board of the Pizza Company of America.

Jack Carter was senior vice president of Pogo Producing Company, an oil and gas exploration and refining firm. He had previously been a managing director of Lehman Brothers, the investment banking firm.

Mike Masson was considered the westerner on the Board. He was president of Sullivan and Masson, Inc., consulting engineers, architects, and construction managers. Earlier, he had been on the board of a savings and loan institution, and had been a past president of the Arizona State Chamber of Commerce.

For all of their eventual differences, this was a relatively homogeneous Board. They were conservative in philosophy, determined to be frugal with taxpayer dollars, and committed to accomplishing a modest program without taking unwarranted technical or financial risk.

ORGANIZATION AND STAFF

The combined action of the two Boards quickly established an organizational structure and recruited capable staff.

The Organizational Approach

Both Boards adopted the same organizational approach, one that reflected the functions that the Corporation had to perform in order to succeed. It had to solicit proposals for financial assistance, evaluate those proposals against competitive criteria, negotiate financial assistance commitments with the most promising project sponsors, and, finally, monitor the sponsor's compliance with the contracts and carry out the government's part of the contracts.

It was evident that the same types of capabilities would be required for each of these organizational functions, but in different mixes. The key capabilities were engineering and technical analysis, financial analysis, and legal. Accordingly, vice presidents were appointed to head the staffs with these capabilities (although the official with that rank for legal services was called general counsel).

In addition, a Projects staff, with two vice presidents, was established to coordinate the work of the functional staffs in producing the necessary documentation and to meet the deadlines fixed by the Board. It was similar to many "matrix" forms of organization. Thus, most staff members reported to their vice president for direction with regard to standards for analytic skills and processes, but to project officers in Projects for preparing timely and integrated products.

This system contained built-in tensions. Occasionally, individuals felt they were serving two masters. But, with few exceptions, the tension was a healthy one in that it ensured that issues did not get glossed over and that thorough analysis was carried out on a reasonably expeditious basis.

Given the analytic requirements of project evaluation, two sub-offices had to be explicitly provided for, namely:

The Environmental and Socioeconomic Analysis Office, which reported to the Vice President for Technology and Engineering. This office was responsible for implementing the considerable responsibilities for these matters contained in the Energy Security Act. For example, Section 131(e) of the Act required that any contract for financial assistance shall require the development of a plan, acceptable to the Board of Directors, for the monitoring of environmental and health related emissions from the construction and operation of the synthetic fuel project. That plan was to be developed after consultation with the Administrator of the Environmental Protection Agency, the Secretary of Energy, and appropriate state agencies. In addition to carrying out specific requirements of the Act, the Office was responsible for evaluating whether or not the projects under evaluation were likely to be able to satisfy all federal, state, and local regulations on the schedules indicated by the sponsors. Finally, the office was also assigned responsibility for monitoring a project's compliance with the act's requirement that projects receiving financial assistance provide for the participation of small and disadvantaged businesses in the construction and operation of the project.

The Cost Analysis Office, which reported to the Vice President for Projects. This office was responsible for analyzing the capital and operating cost estimates of project proposals to determine the likely accuracy of such estimates and to determine how much uncertainty to include in the financial analysis of project viability, and, ultimately how much contingency to include in the financial assistance packages under negotiation. In the Corporation's eyes, it was essential that the financial assistance be adequate to carry projects safely through periods of potential difficulty—e.g., to cope with delayed completion of construction, or lengthy periods of getting the technology to function reliably at close to design production levels.

Figure V-1 portrays the organizational structure of the Corporation.

Staffing

Both of the first Boards were cognizant of the difficult and complex negotiations that would take place in structuring financial deals that could entail multi-billions of dollars. They were determined

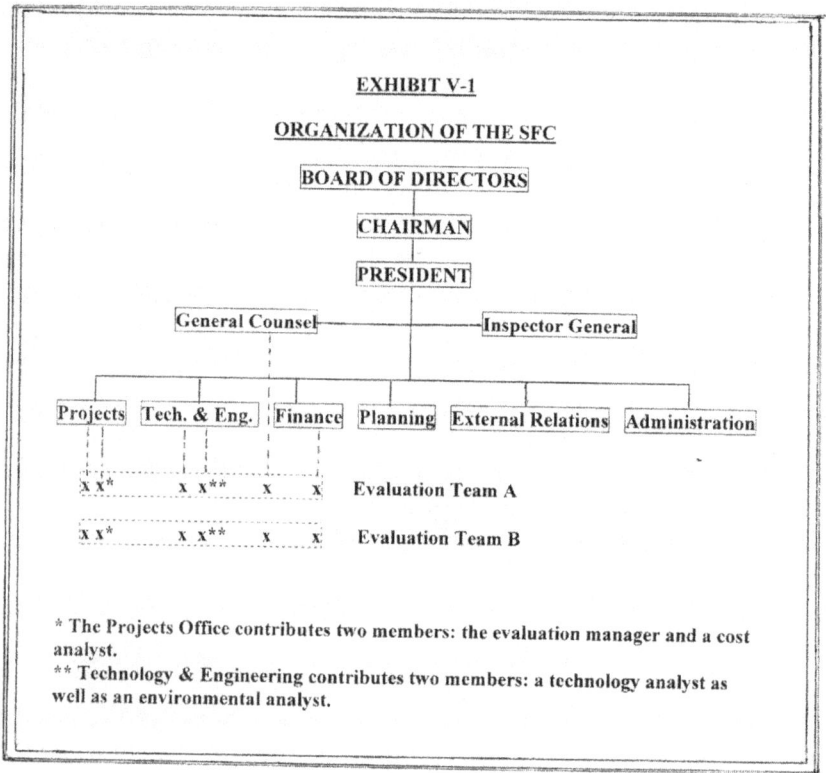

EXHIBIT V-1

ORGANIZATION OF THE SFC

BOARD OF DIRECTORS

CHAIRMAN

PRESIDENT

General Counsel ———————— Inspector General

| Projects | Tech. & Eng. | Finance | Planning | External Relations | Administration |

X X*   X X**   X   X   Evaluation Team A

X X*   X X**   X   X   Evaluation Team B

* The Projects Office contributes two members: the evaluation manager and a cost analyst.
** Technology & Engineering contributes two members: a technology analyst as well as an environmental analyst.

that the Corporation would have the expertise necessary to protect the government's interests. In this, they drew on the flexibility offered by the Corporation's quasi-federal structure and the latitude regarding salaries offered by the founding legislation, particularly for staff in financial analysis and legal affairs recruited from the private sector. Recruiting such individuals from the private sector ensured that the Corporation would had the expertise to negotiate complex financial contracts potentially involving billions of dollars with major private corporations and their law firms. Otherwise, the Corporation recruited from a variety of sources for its staff—looking both to the private sector and to government.

To illustrate the caliber of the recruited staff, the background of a number of officers of the Corporation was as follows:

Leonard Axelrod, the only Vice President for Technology and Engineering during the Corporation's existence, was a retired officer with substantial R&D experience from Kellogg-Rust, one of the nation's premier engineering firms involved in the construction of new technologies in the petroleum and process industries.

Jimmy Bowden, who was the Corporation's first Executive Vice President, had been an officer with Conoco.

Ed Miller, who was one of the first Vice Presidents for Finance at the SFC, had previously been a financial officer at AMAX, the large minerals firm.

Hod Thornber, who succeeded him as Vice President for Finance, went on to become a Managing Partner of Putnam, Hayes and Barlett, the financial consulting firm.

Len Rawicz had been Deputy General Counsel of the Department of Energy, before becoming General Counsel of the SFC. He went on to become of Counsel to Skadden, Arps, one of the nation's largest and most prestigious law firms.

Dwight Ink was Vice President of Administration of the SFC for a number of years. He had had wide experience throughout the federal government, heading a number of agencies, before coming to the Corporation.

There were other capable individuals as well, generally for briefer periods. Ed Cox and Jeff Lipkin also served as general counsels. Jim Groelinger was the financial vice president in the Corporation's last period. Bob Fairman also served as a Vice President for Administration. As someone with experience at the Department of Energy in leading project evaluations for the fast track pre-Corporation program, I served as the Vice President for Projects, and Charles Cowan was Senior Vice President for Projects.

As is evident from the above list, some turnover took place in the Corporation's officer ranks. Although congressional critics tried to exploit this phenomenon as a sign of SFC management problems, a more accurate assessment would be that the turnover was consistent with the Corporation being a short-lived institution. Some individuals intended to join the Corporation for only a couple of years or so. But the Corporation could provide a career for no one. A number of the staff made their contributions and then moved up the career ladder elsewhere as partners in law firms, partners in financial consulting firms, senior officers in banks, heads of agencies, and so forth.

THE SOLICITATION PROCESS

The program was authorized to provide very large amounts of financial assistance to private firms, up to $3 billion could be provided to a single project. How could the Corporation protect taxpayer funds from being wasted on unworthy projects—or even just projects whose connections were sounder than their designs?

Provisions of the Act

The Energy Security Act itself tried to provide safeguards to help ensure that worthy projects were selected to receive assistance. Particularly, Section 131 stated "Financial assistance shall be awarded to a qualified concern whose proposal is most responsive to a solicitation for proposals issued under the authority of Section 127 and is most likely to advance the purposes of this title, including consideration of price and other factors. Whenever, in the judgment of the Board of Directors, it is practical and provident to do so, the Corporation shall award financial assistance on the basis of competitive bids." Section 127 provided the authority for the Corporation to issue solicitations from time to time and established some procedural prerequisites for the issuance to follow.

While the Act went on to specify a few of the criteria to be employed in the selection of projects (which will be discussed in the next major section of this Chapter), the broad design parameters were left up to the Board. Some of the wide latitude permitted was evident in the divergent approaches taken by the Sawhill and the Noble Boards.

The Initial Solicitation

Section 127(a)(3) of the Act directed the Corporation to issue its initial set of solicitations before December 31, 1980. Inasmuch as the Board was not appointed until October of that year and its essential first order of business was to approve an organizational structure and hire a cadre of staff, time was short. Consequently, the Chairman, in line with the congressional desires to move swiftly, adopted a minimalist approach to the Corporation's initial solicitation.

As a result, the initial solicitation consisted only of three pages! If this strikes the reader as unremarkable, consider that it is not uncommon for a government solicitation to run to fifty or more pages (with the complexity to match). The solicitation merely indicated the nature of the program being pursued, the Corporation's desire to receive proposals by March 31, 1981, and principal selection criteria.

The Board's philosophy, in part, was that only the most sophisticated and well-heeled corporate entities had any hope of raising the hundreds of millions of dollars of private equity required by the Act or of managing the construction and operation of the resulting pioneer plants. Consequently, the thinking went, these groups well knew the considerations that entered their own decisions for supporting—or not supporting—a project, and would present the commensurate information to the Corporation with the proposals.

Then, the Board would be in a position to identify the projects of greatest interest and could elicit additional information in a more detailed (and costly) Phase I proposal. The solicitation also referred to statutory criteria for negotiating terms of assistance—i.e., a competitive selection would be made that would produce a diversity of technologies for either the least amount of obligational authority for an individual facility or the least amount of assistance per barrel produced. (These are usually contradictory given the impact of economies of scale, but are nonetheless useful for the selection of projects.) Thereby, everyone would save money. Sponsors would not have to invest hundreds of thousands in proposals to determine if the government was really interested in the project, and the Corporation's staff would not have to invest inordinate amounts of time in reviewing lengthy proposals for projects that were of limited interest. The process approximated what the private sector might employ for itself.

It is conceivable that such a streamlined process could work if the Board were willing to make the sharp decisions to eliminate projects on the limited amount of data presented and be willing to defend these decisions as competitive and fair as defined in the Act. Such decisions might have to be defended both in courts as well as in front of Congress.

In the event, the process never unfolded that far. The Sawhill Board had "resigned" before the first proposals had been received. Sixty-eight proposals were received by the closing date, which was during the interim between Boards (see Appendix B—Solicitation Summary). The new Board, in one of its first decisions, decided to tighten the solicitation requirements at the recommendation of the staff.

The problem that had arisen was that proposals were not submitted only by responsible parties. A few were comprehensive and were well supported by engineering data. These tended to be the projects that had received government assistance for feasibility studies or cooperative agreements under the interim program run by the Department of Energy, which had been tailored just for the purpose of getting projects ready for serious consideration by the Corporation. A number of other projects looked promising, but lacked supporting evidence and detail. Even responsible parties were not willing to invest over much in such proposals until they had gained the measure of the seriousness of the Corporation, especially in light of the perceived lack of enthusiasm of the new Reagan administration for synthetic fuels.

Still other proposals were skeletal outlines of potential projects. It was clear that the sponsors had not done much serious work and were merely trying to get in the queue for eligibility to be considered for assistance.

And, inevitably, there were opportunists. The Corporation received several proposals from individuals who insisted without any tangible evidence, that they had invented a process that would produce fuels at competitive prices and supposedly merely needed the ability to raise some capital, which they were unable to get from the private sector (for good reason given the lack of evidence to support their claims). In a few instances, they were cranks who bombarded the Board and the staff with letters, telegrams, and telephone calls. Some individuals did this over a several year period. One of the more extreme proposed to tow icebergs to Virginia to provide water to the project. One even gulled ABC, which enthusiastically presented the claims on the show "Isn't it Amazing." That promoter was eventually sent to prison for mail fraud.

The above classes of proposals were not difficult to distinguish from one another once requirements were established for evidence that verified claims of technical performance. The class of proposals that proved more difficult to evaluate came from entrepreneurs who seemed quite honest, but who, because of limited financial backing, had undertaken very little project-specific design work. The cost of the minimal amount of work necessary to support a proposal and provide a sound cost estimate could entail some millions of dollars amounting to several percent of the final project cost. These sponsors usually had the good sense to propose projects with relatively proven technologies so that they did not have to support performance claims. Some would propose generally sensible—even innovative—project designs. Nonetheless, they tended to be deficient in financial backing and had not done the work to provide a reasonable cost estimate. In effect, they did not yet have the equity to develop a project even if the Corporation were to have provided assistance. Why, then, did they proceed as far as they did? They hoped to negotiate an attractive amount of assistance from the Corporation in a deal they could then market to those that did have the equity, keeping the entrepreneurs' percentage of project ownership for themselves without having to put up a comparable share of capital.

In this group, there were promising projects were that had done only modest amounts of design work and lacked sufficient equity for the project to proceed, but did already have one or more financially substantial private firms in the sponsoring group. This class of projects tended to provide most of the projects that ultimately were endorsed by the Board for assistance by approving letters of intent (to be discussed later).

Finally, there were proposals from well-heeled sponsors who were fully capable of raising the required equity and in managing the construction and operation of the resulting projects all on their own. This class of projects was unfortunately the smallest in number.

The primary difficulty faced by the staff was how to focus on the subset of the 68 proposals that were of real interest. First of all, it was not always possible to tell definitively from the sparse information submitted by most sponsors. Then, even if one determined which of the projects were worthy of further investigation, how could the others be eliminated from consideration and still remain within the parameters of what the Act termed open and fair competition among proposals. As crafted, the Initial Solicitation did not contain much in the way of explicit criteria for action. An even greater fairness issue arose with the recognition that virtually all the projects would require further development before an agreement could be reached. Accordingly, substantially more proposal information would have to be submitted by those projects before they reached the final hurdle. The crux of the issue was: if a few projects were given the opportunity to continue development to the point they were attractive to the Board for signing a final commitment, why should not all? Conversely, if all could do so, the field of competition would remain so large, that few projects would find it in their interests to make the large investments that would be needed. Moreover, the Corporation's staff would have to be large to deal on such a basis with a large number of projects.

In any event, after nine months or so, the Corporation had a Board compatible with the new Administration, had an organizational structure along with officers and staff, and was ready to tackle the large number of projects that had expressed interest in building the new synthetic fuels industry. The next order of business was to redesign the solicitation process to incorporate the lessons learned to date. Appendix C provides a timeline to assist the reader in tracking the multiple solicitations eventually issued by the SFC as well as the key project evaluation and selection points.

VI

Charting the Mission

With the Corporation operational, the new Board under Chairman Noble had to decide specifically how to deal with the 68 proposals received under the initial solicitation and more generally how to design an ongoing solicitation approach for attracting projects that would expeditiously, but realistically, meet the objectives of the Energy Security Act. The objectives were clear enough, but how should the authorities provided by the Act be deployed in light of the legislation's directive to award assistance on the basis of *competitive* bids considering unit production costs of the synthetic fuel, the varying status of project development across the different energy sectors, and the demanding nature of new technology development itself?

In this light, the Board had to consider a number of key issues: how to winnow clearly unattractive or unresponsive proposals on a fair basis, how to give guidance to the others regarding the provision of additional information that would allow the Board to meet the requirements of the Act regarding competitive costs, and how to encourage the most attractive but not yet ready projects to continue their development.

The following sections lay out the considerations underlying the guidance that the Board gave to the first group of applicants and the structuring of the comprehensive solicitation process that governed the Corporation's operations throughout its existence. In this, the Corporation strove to ground the process in an empirical understanding of new technology development, especially on the central importance of project sponsors demonstrating technical readiness and being able to estimate accurately the capital and operating costs of proposed commercial-scale facilities.

Need for project definition

It was evident to the Board that making informed decisions would require them to know far more about the projects being proposed than was contained in the initial proposals, especially about the proposals' technological readiness, the quality of underlying engineering design work, and the prospects for the sponsors to raise their share of the capital. Without such information and an accurate estimate of the cost of facilities, the Corporation would have found it difficult to select projects competitively, considering both relative unit costs of production as well as amounts of assistance. Particularly in the case of loan guarantee assistance, if a project significantly overran costs and thereby exhausted the amount of loan guaranteed, the chances were that the project would fail. To be sure, Congress had permitted the Corporation to renegotiate assistance contracts under some such circumstances. As a practical matter, however, all the contracts negotiated were on the basis that there would be no additional assistance. It could hardly be otherwise: how could an award be made to one project on the grounds of being the most economically competitive, if it was to receive more assistance later?

The Board's quandary was that it would cost sponsors millions of dollars to bring projects to the point that such questions could be adequately addressed. But sponsors were not likely to invest the necessary time and money until they knew whether the new Board was serious about prosecuting the program and unless they had a way of judging their prospects. So, the Board had to develop a solicitation process that met the needs of both the Corporation and the sponsors while showing Congress that the prospects for assembling a stable of promising synthetic fuels projects were good.

In any event, there was no escaping the fact that a minimum level of project definition and design would be essential to allow an informed and competitive selection process. How minimum? What information was essential?

The nature of the information needed and reasonable for the Corporation to require was highlighted by a groundbreaking study that had been commissioned by the Department of Energy and was just being completed by the Rand Corporation. The first half had been completed in 1978, while the second was in draft form as the Corporation was being formed and was familiar to much of the staff.[64] Indeed, a chief objective of the Rand work was to provide an experiential framework to assist potential decision-makers in using cost estimates effectively at relatively early project stages. More specifically, it was to seek a better understanding of the reasons for inaccurate estimates of capital costs and performance difficulties for first-of-a-kind process plants, especially energy process plants.

Then, armed with a better understanding of the problems, government and industry, and in this instance the Corporation, would have tools to improve assessment of the commercial prospects of pioneer technologies. The Report's conclusions, based on an analysis of 44 pioneer process plants built by the private sector over the prior fifteen years, were very cautionary about pitfalls in cost estimating that await those planning new process plants, especially those that involved solids handling, as did most of the potential synthetic fuels facilities.

Estimating Capital Costs

The Rand work made more tangible the conventional wisdom that the ability to estimate the costs of a new facility is a function of how well defined its physical and chemical processes are as well as its engineering plant design. Analysts have to ask whether similar facilities have been built and operated before, whether the technology has been operated at design-scale before, whether construction plans are site-specific, and whether construction plans have been drawn up sufficiently to price out materials and labor. Costs can be known with considerable accuracy after all the detailed engineering is complete and all the drawings and equipment lists are available for vendor quotes, material takeoffs, and labor estimation. Such estimates, of course, contain no actual knowledge of construction bottlenecks that might arise, of system redesign made after the beginning of construction, or of inflation. But such things occur on all projects to some degree and allowances can be made for them.

It is, however, an expensive proposition to undertake the design work required to get a cost estimate with a narrow uncertainty

Average length (in months)	25	14	20	24	7
Range	0–170	0–54	4–57	4–53	1–34

R&D	Project definition	Engineering	Construction	Start-up

| Estimate classes | 0 | 1 | 2 | 3 | 4 |

Figure VI – 1 — Project stages and duration for sampled plants

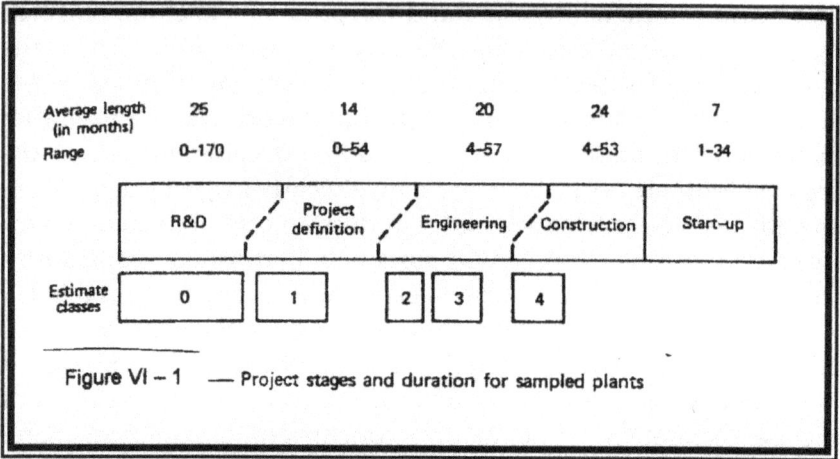

band. If a decision to undertake a plant has already been made, these estimates emerge as a matter of course. But what if the decision to proceed depends on the cost? How much should one be willing to invest in advance of knowing the answer, assuming that the project might or might not go ahead?

The answer is to draw on experience in building other first-of-a-kind facilities in comparable industries. The Rand Study established an overview of project development stages that could be employed to consider such questions. Figure VI-1 illustrates how the study depicted a project's development in five stages: R&D, Project Definition, Engineering, Construction, and Start-Up.[65] The Figure also indicates the points at which cost estimates are typically made, and labels them from 0 through 4.

Let's examine the characteristics and accuracy of cost estimates associated with each. Beginning with the first stage, as noted by Rand, the function of R&D is to provide the basic theoretical understanding of the process necessary for establishing its feasibility or to work on some particularly troublesome aspect of the plant. If the process is new, R&D may include a process development unit (a small batch unit that tests part of the process). This scale is almost always a fraction of that employed for a commercial production facility. Typically, in order for the synthetic fuels plants to be considered by the Corporation, a scale-up of ten or more to one from the largest test facility to the proposed production facility was involved. Industry tends to be skeptical of cost estimates developed at this stage because relatively little is specified about the full-scale plant. Moreover, most R&D performers have little experience with plant design and construction.

Analysis of a commercial plant begins with the project definition stage, which entails defining the scope of the proposed plant, the basic plant layout, and the process flow conditions. Most major equipment needs are also defined at this time and examination of a possible site is begun. The amount of effort required at this stage can vary considerably depending on past experience with the kind of unit involved, and how much is already known about the proposed site. If no project has been previously built at the proposed scale, which was largely the case for synthetic fuel plants, data from an integrated pilot plant is crucial. If no such facility exists, one might have to be built at considerable cost. If a comparable facility already exists, tests on the proposed feedstocks can provide a lot of design information. For example, the resulting data provide key inputs for determining heat and material balances—i.e., input-output equations modeling the flows of energy and process materials for each unit of the plant. The balances govern the sizing of equipment. An important product of this stage is a preliminary cost estimate, called Class 2 by Rand.

The next stage is Engineering Design, which is the process of turning preliminary drawings and specifications into a "blueprint" from which the operating plant can be constructed. Procurement specifications and detailed phasing for construction are developed. When engineering design is 30 percent or more complete, a new cost estimate (Class 3) is usually made, often called the budget estimate. This estimate is used as the basis for planning capital expenditures and to establish the cost accounts that will be used to control expenditures during procurement and construction. This estimate is often the last estimate formally presented to management for authorization.

When engineering is nearly completed, a final "definitive" estimate (Class 4) is sometimes made. Since data for this estimate apply to a fully designed plant with firm bids on equipment and subcontracts, the confidence in this cost estimate is within about five percent.

The beginning of the Construction Phase varies from plant to plant, but typically is when the design is less than half completed. The construction period ranges from less than one year for small plants to more than four years for large complexes. The commonest causes of delay are late delivery of materials, unavailability of labor, poor productivity, inclement weather, and strikes. Only rarely is construction delayed by errors in plant design. Usually, design problems will not appear until plant startup, which Stage, usually requiring about one to six months, begins as soon as construction is completed.

Working within this schema, the Rand study drew on a database of 44 commercial-scale chemical process plants, which were selected to shed light on energy process plans—a large enough database to sustain a conclusive analysis of cost estimation and performance problems of similar plants.[66] The Rand study found that cost estimates for new process technology plants deviated widely from the actual costs of the constructed facility. More importantly, the reasons for cost growth were not generally well understood, or at least not explicitly articulated by those in the estimating business. Conventionally, it was expected that estimates would peak modally around actual costs with a certain standard deviation; and the earlier a project was in project development, the greater should be the deviation. Examining the various estimates made for the projects in their study, Rand found a somewhat different picture. Figure VI-2 presents a schematic "Experience of the pioneer plants sample with estimation accuracy".[67]

One sees that early estimates tend to be much too low, but estimates do improve with time and effort, and consequently standard deviations narrow. Note that earliest class of estimates averaged less than one half the eventual plant cost and, in some instances, reflected less than one-third the eventual cost. Why is that? The Rand authors began with factors cited most often in the literature and attempted to determine their statistical significance in explaining the differences between estimates and actual costs. The three groups of causes examined were (1) project uncertainty and estimation methodology, (2) process uncertainty, and (3) external effects on cost (e.g., inflation, bad weather, regulatory changes, and strikes).

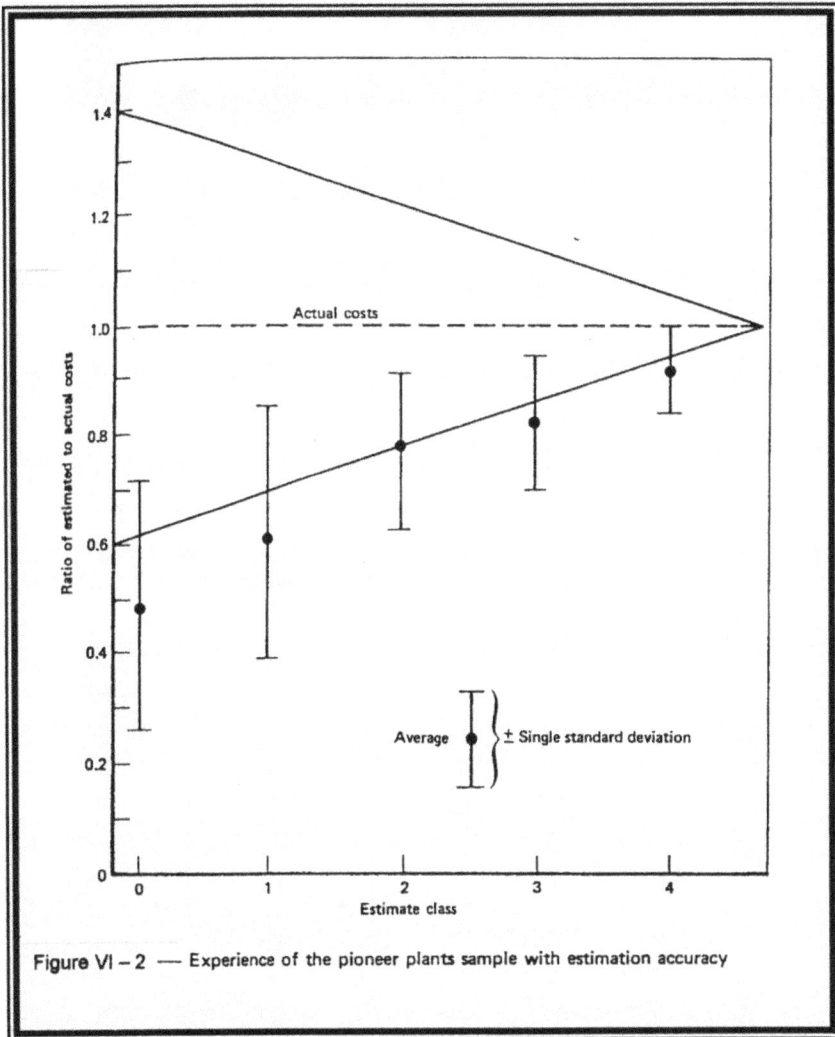

Figure VI – 2 — Experience of the pioneer plants sample with estimation accuracy

At first glance one is tempted to emphasize the latter. Rand, however, found that not to be the case. As is evident in Figure VI-3, entitled "Importance of cost growth vs. external factors in underestimation of costs," only about 26 percent of the misestimation of plant costs could be attributed to such external factors.[68]

The greatest portion of misestimation was rather attributable to the *cost growth* of the projects themselves. The cost growth is the result of the lack of knowledge about the facility being estimated.

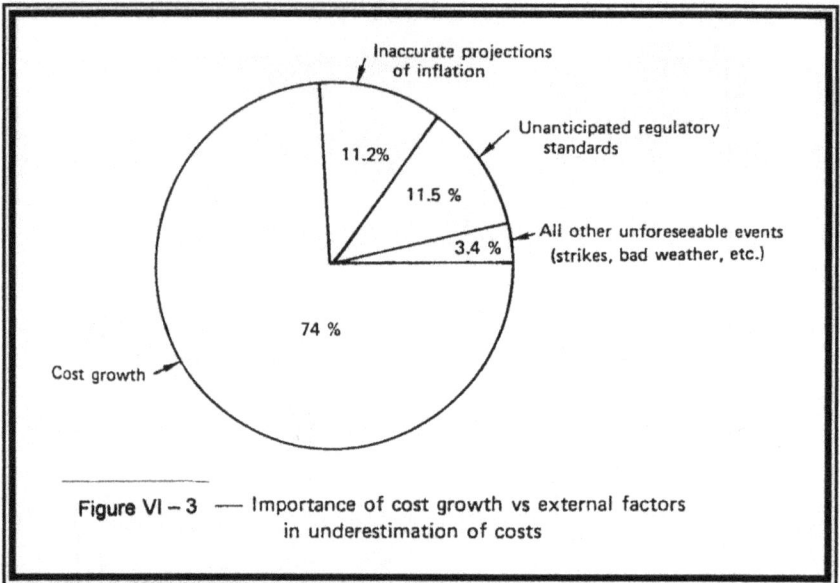

Figure VI – 3 — Importance of cost growth vs external factors
in underestimation of costs

As stated by Rand,

> Engineering estimation methods, no matter how carefully devised, cannot fully compensate for scantiness of information. Particularly when little project definition has been accomplished, and commercially unproven technology is to be employed, the estimator has no relevant prior experience with the same site and process configuration to guide his estimates. Cost growth stems primarily from the fact that at early stages of engineering, especially for pioneer plants, many cost elements cannot be estimated because they simply do not yet exist.

This phenomenon was captured in the Table VI-1, "Cost-growth Ratios by Class of Estimate for Pioneer Plants Study Sample," which shows that estimates based on detailed engineering (Class 4) are within 7 percent of plant actual costs, but estimates based on conceptual designs during R&D (Class 0) were less than half of actual plant costs.

Cost estimators have long been aware of this type of relationship, and have attempted to provide for escalation in early esti-

COST-GROWTH RATIOS BY CLASS OF ESTIMATE FOR PIONEER PLANTS
STUDY SAMPLE

Class of Estimate	Average*	Standard Deviation	Number of Estimates
0	0.49	0.23	7
1	0.62	0.23	18
2	0.78	0.15	30
3	0.83	0.12	27
4	0.93	0.08	24

*Ratio of estimated to actual costs (excluding external factors).

Table VI-1

mates by setting aside larger contingencies to cover the uncertainties. The difficulty, however, is not just the fact that a given process
is unproven. Instead, it is more the unforeseen design, engineering,
construction, or start-up problems that unproven technologies can
run into and that often require expensive redesign or repair. Such
problems were likely to be a function of new technology being employed in a plant as evident in: percent of capital cost in technology
unproven in commercial use; number of steps (counted on block
basis) incorporating new technology; number of integrations of
proven steps that had not been integrated in commercial use before; the percentage of the heat and mass balance equations based
on actual data from prior plants rather than calculated from theory;
and whether this was the first time that a technology had been used
commercially in the United States or Canada.

Rand found that this model was highly accurate for assessing
cost uncertainty. Nearly one-half of the projects' cost inaccuracies
were predicted within plus or minus five percentage points. Virtually all were predicted within plus or minus 15 percentage points.

Estimating Economics of Process Plant Performance

While capital costs tend to receive most estimating attention, costs
associated with plant performance can be just as influential in establishing plant economics. For example, if a plant operates at only

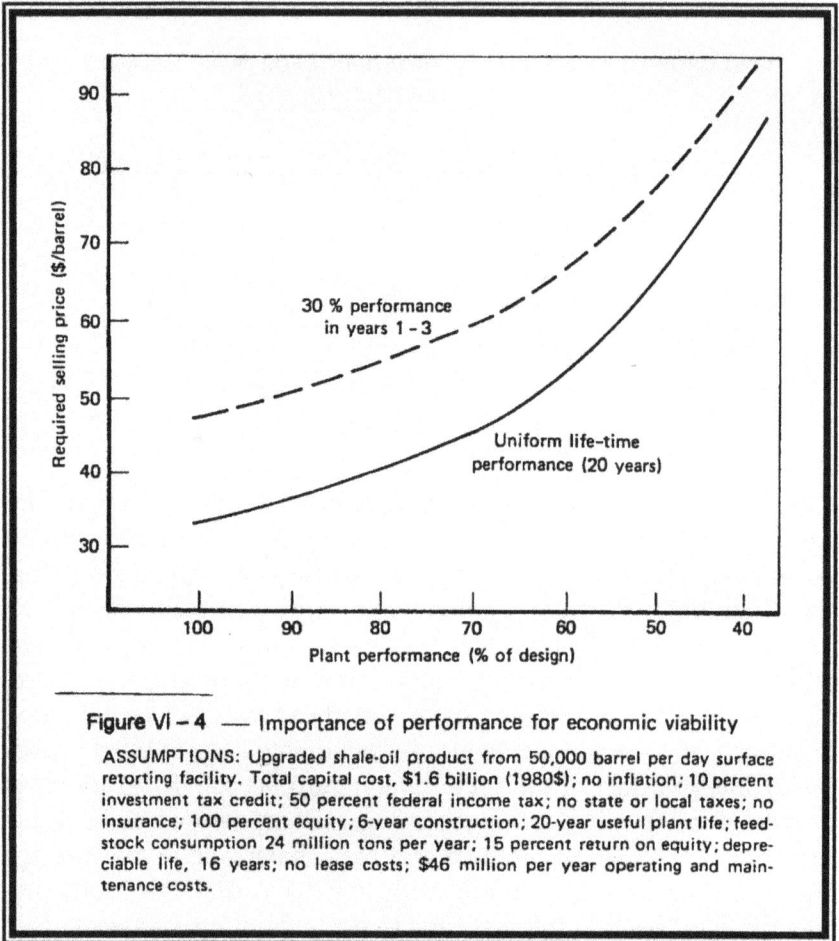

Figure VI – 4 —— Importance of performance for economic viability

ASSUMPTIONS: Upgraded shale-oil product from 50,000 barrel per day surface retorting facility. Total capital cost, $1.6 billion (1980$); no inflation; 10 percent investment tax credit; 50 percent federal income tax; no state or local taxes; no insurance; 100 percent equity; 6-year construction; 20-year useful plant life; feed-stock consumption 24 million tons per year; 15 percent return on equity; depreciable life, 16 years; no lease costs; $46 million per year operating and maintenance costs.

fifty percent of nameplate capacity, the effective capital cost per unit of output is nearly doubled, and in some cases the operating and maintenance costs may more than double. Further, if design problems substantially delay the operation of the facility, cash flow and the resulting rate of return can be sharply reduced. This is seen in Figure VI-4, "Importance of performance for economic viability"[70].

The Figure illustrates that in the example of a shale plant, if performance is at only a 30 percent level for the first three years of a plants life, the selling price of a barrel of shale oil (or level of price guarantee negotiated by the Corporation) would have to be roughly fifty percent higher for the plant owner to derive his projected rate of return than would be the case if the plant operated at capacity after a brief startup period.

In the Rand study, startup times for process plants ranged from as little as one month to as much as six months for large and complex units. The planned startup period averaged three and a half months. In actuality, about half of the plants failed to reach the usual goal of 85 percent of design capacity in the second six months after start-up. Indeed, about 23 percent of the plants failed to reach 50 percent of capacity in the first year. Many of those units were later permanently de-rated and firms reported having lost money on most of those plants.

Rand attempted to isolate those factors that appeared most to explain poor plant performance. The Report found that four variables could account for 90 percent of the actual performance:

The number of steps new in commercial use

The percentage of the heat and mass balance equations based on actual data from prior plants

The level of design difficulty encountered with waste handling

Whether the plant processed solids.

The closing words of the Rand report were prophetic:

Because the costs of these technologies will almost assuredly be higher than expected and the actual output lower, the near-term prospects for major increases in liquid hydrocarbon supplies from energy process plants are not good. But the longer-run benefits of having designed, constructed, and operated a well-chosen set of pioneer synthetic fuels plants could prove to be a national blessing in the 1990s.

A careful reading of the Rand Report could only be sobering for those beginning to implement the ambitious goals of the Energy Security Act. As the reader will recall, Congress had called for national synthetic fuels production of 500,000 barrels a day by 1987—just seven years after the passage of the Act. But Congress seemed to accept such uncertainties and had directed the Corporation to get on with the implementation of the Energy Security Act. How could it best accomplish that goal?

Establishing selection criteria

Armed with this conceptual framework, the Board decided that projects should be required to achieve certain levels of design maturity before amounts of financial assistance would be negotiated. They did so even though few projects had yet reached that stage. Those that were ready tended to be ones that received earlier design and planning grants from the Department of Energy. (Public Law 96-126, which authorized funds for the Corporation, also provided authority for grants to be made to potential proposers for the design work necessary for feasibility studies and early engineering of projects before proposing to the Corporation.)

The Corporation's new course of demanding a minimum level of design maturity and financial backing clearly imposed a significant hurdle for applicants to surmount. This had the obvious advantage of focusing on projects having the greatest chance of success at least cost to the taxpayer. On the other hand, such criteria would significantly limit the universe of potential applicants; indeed, most of the initial applicants either chose not to resubmit or would not make it through the first screening stages of the new solicitation approach.

The selection criteria were that the project would produce a fuel eligible under the Act, that the project was mature in terms of having accomplished a preliminary array of developmental work, and that the project had already completed a defined amount of engineering design work along with cost estimates, and be able to complete a more detailed cost estimate before negotiations for financial assistance could commence. Let's examine these more closely.

The tests for a mature project were that:

Rights to the plant site had been secured;
Comprehensive financial and technical feasibility studies had been completed;
Sufficient design work had been completed to allow costs to be estimated with a high degree of accuracy ;
The necessary resource, utilities, and water were accessible;
Rights to key technology were assured;
Estimated outline of amounts and terms of commitments of each sponsor had been completed; and

Permits could be obtained consistent with the construction schedule.

At a later time, these criteria were expanded to require:

Active involvement in the project of at least one economically substantial sponsor that was a "qualified concern" within the meaning of the Act;

Assignment to the project of capable personnel actively functioning in key management positions;

Completion of a defined level of design; and

Commitment of financial resources.

Once a project's maturity was ascertained, it was to undergo an assessment of its strength and value to the program in meeting the goals of the Act. Strength evaluation was concerned with: management capability, technology and engineering status, product marketability, availability of feedstock and utilities, amount and form of financial assistance requested, project economics, sponsor financial strength, and ability to meet environmental, socioeconomic, and water requirements. The status of engineering design and cost estimates were defined according to the convention shown in the Definition of Design Levels below.

Definition of Design Levels

Level II design is characterized by:
Preliminary process flow diagrams and heat and material
    balances
Preliminary piping and instrumentation
Preliminary major equipment list and sizes
Preliminary plot plan
Preliminary Project schedule
Equipment costs from vendors and in-house estimates
Total installed cost from discrete cost factors

Level III design is characterized by:
Final process flow diagrams
Detailed piping and instrumentation flow diagrams includ-
    ing utilities

Equipment sizes with specifications and materials
Overall plot plan and site characteristics
Electrical single lines
Elevation drawings (critical areas)
Project schedule based on quoted deliveries and site conditions
Bulk materials from preliminary takeoffs and labor derived by manhours and rates
Total installed cost from materials/equipment quotes plus construction estimates to construct

Level II/III design means that the design is of at least Level III for units containing key conversion technologies, process units using technologies not previously demonstrated in commercial operations and process and offsite units not readily extrapolated from previous experience with the balance being at least Level II design.
Note: The definitions are modified for tar sands/ oil in situ extraction projects.

In developing the above evaluation criteria, the Corporation strove to adhere to private sector practices that would in most instances not cause capable sponsors to incur unnecessary expenses or depart from the prudent development approaches they would employ on their own.

As the criteria were fixed, each of the Corporation's functional offices devised methodologies for evaluating the project proposals against these criteria. Thus, for example, the vice president for finance determined how to assess a project's likelihood of long-term financial viability and for assessing the likelihood of a given group of sponsors for raising the necessary amounts of equity. This required in turn the development of a financial computer model targeted to synthetic fuels type of facilities.

A REVISED SOLICITATION PROCESS

The Board drew on the above insights as well as the results of reviewing the proposals received under the initial solicitation to develop a comprehensive solicitation process.

As a first order of business in December 1981, the Board established a new deadline of January 1982, which was two months hence, for the sixty-eight proposal sponsors to submit additional information if they wished the proposals to remain under consideration and included specific criteria for eliminating unresponsive proposals[71]. In the form of a Supplement to the Initial Solicitation, the Corporation identified the additional information required of project sponsors, provided an overview of evaluation criteria adopted by the Corporation, and explained more fully how the competition would be conducted.

A few projects were already far enough advanced that they could readily provide the additional information. These were projects that had received assistance in the form of feasibility study grants and cooperative agreements from the Department of Energy's fast start program. Consequently, they would be the first out of the gate. Other projects were given the chance to apply to one of the new general solicitations—providing they were willing to make the necessary investment in plant definition and design. The Corporation formulated a comprehensive solicitation process that provided for several additional solicitations at staggered intervals, as illustrated in Figure VI-5.[72]

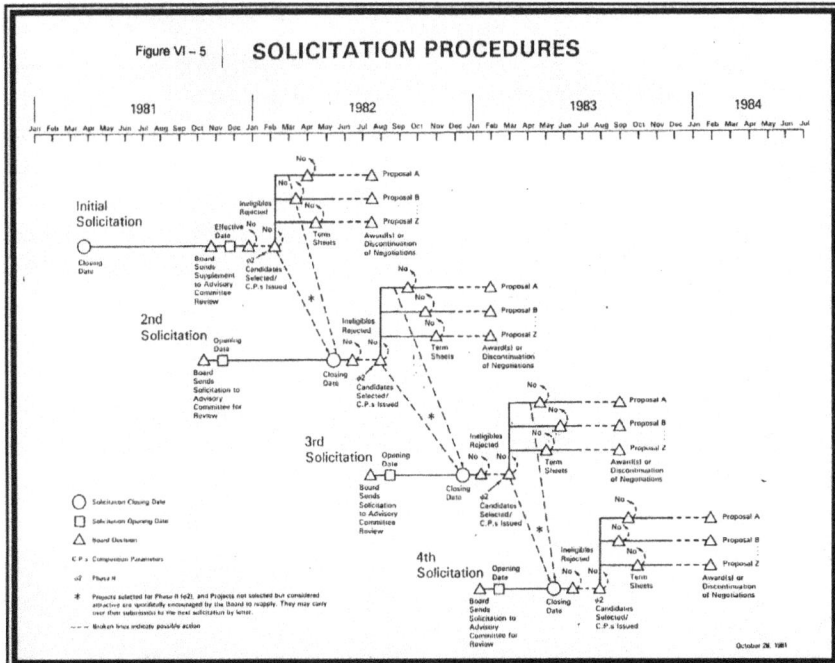

Figure VI – 5 | SOLICITATION PROCEDURES

The new solicitation process was designed specifically to recognize that not all projects were equally mature. It was clear that most projects would not be prepared to submit responsive proposals for many months or even a year. The staggered schedule allowed sponsors to choose the best time to submit a proposal, and they knew that if unsuccessful under one solicitation they could reapply under a subsequent one. By providing more clarity on selection criteria and by providing multiple opportunities to apply, the Corporation hoped to provide an incentive for interested parties to continue to make the investments necessary to improve the design maturity of their projects.

The new procedures were also designed to nurture project development during the review process by providing for a separate Phase I and Phase II review. In Phase I, the projects were reviewed against the maturity and strength criteria discussed previously. In these, the Corporation sought to determine that sufficient operating experience existed with the proposed technologies to judge that the project had good prospects for operating success. In addition, an approximate cost estimate based on site-specific factors had to be in hand indicating some prospects for economically viable operation over a normal commercial period—with SFC assistance, of course. Finally, the proposal had to demonstrate that existing project sponsors had the likelihood of completing materials for Stage II consideration, and for being the nucleus of a completed sponsoring group that could satisfactorily build and operate the facility.

If a project met those criteria or demonstrated that it could meet them within a reasonable period of time, the Board would advance the project to Phase II. At that point, a project would submit a more detailed proposal based on more engineering design and on a cost estimate having less uncertainty. The Corporation staff would conduct a deeper review that included verification of information received from the sponsors, such as an independent cost estimate for projects requesting a loan guarantee and an independent review of the proposed resource base and water supplies. It was made clear to sponsors that this was not a typical government procurement, which minimizes contact with sponsors after proposals are submitted. Indeed, the Corporation wished to nurture the projects along during the evaluation process by providing for information exchanges, supplemental submissions, meetings, reviews, and site visits.

Concurrent with the detailed evaluation, the Corporation negotiated with project sponsors the basic business terms of possible financial assistance. In numerous cases, these negotiations led to the Board approving non-binding letters of intent, a step short of an assistance agreement. The letters proved a useful tool for enabling projects to attract additional investors. A discussion of such letters and their uses will be provided in the next chapters.

VII

Shakedown of Operations and the First Award

With a new Board in place, on February 8, 1982, the President signed Executive Order 12346 declaring the Corporation to be operational. Thereby, unused Defense Production Act funds were transferred to the SFC from DOE as was the administration of the Parachute Creek and Colony financial assistance agreements that had been negotiated by DOE during the head start program included in the Energy Security Act. And with the mission charted and the solicitation tools in hand, the Corporation moved expeditiously to attract and evaluate an array of diverse qualified projects. Over the next year and a half, it completed work on the Initial Solicitation, issued the next two general solicitations, evaluated the proposals received, signed the first letters of intent to issue awards, and completed the first financial assistance contract.

Initial solicitation efforts

While only one of the contenders from the Initial Solicitation ultimately received assistance from the Corporation, a larger number did receive letters of intent from the chairman in which he indicated that the corporation would complete an assistance agreement if the sponsors met certain final conditions. Examining the staff's efforts to evaluate and negotiate with the contenders from this solicitation provides a dynamic view of how the Corporation's staff and procedures gelled early on to support the overall mission.

As noted in the previous chapter, the new Board required the 68 responders to the Initial Solicitation to provide substantial additional information to the Corporation if they wished their proposals to be considered further. This information had to be received within two months of the Board's guidance—that is, by January 4, 1982. Only 28 of the original proposals met the deadline.[73] Accordingly, they were the only ones to be considered further in the three evaluation stages: the maturity review, the Phase I strength evaluation, and the Phase II strength evaluation.

The First Maturity Review

As a matter of course, the first review of proposals involved a check of eligibility against the criteria specified under the Act. In all but a few instances, this review could be done by immediate inspection, and consequently it did not constitute a separate phase of evaluation. In terms of serious effort by the Corporation's staff, the maturity review constituted the first hurdle encountered by proposers. This hurdle consisted whether the proposal could demonstrate that:

Rights to the plant site had been secured;

Comprehensive financial and technical feasibility studies had been completed;

Sufficient design work had been completed to allow costs to be estimated with a high degree of accuracy during the strength evaluation;

The necessary resource, utilities, and water were accessible;

Rights to key technologies were assured; estimated outline of the amounts and terms of commitments of each sponsor were completed; and

Permits could be obtained consistent with the construction schedule.

Of the 28 projects that submitted additional required information by the January deadline, only eleven were judged to meet the maturity criteria and were advanced to the strength evaluation.[74] Let us examine these projects, effectively the most advanced synthetic fuels projects in the United States at the time, individually.

The Hampshire Energy Project, which was to be located at Gillette, Wyoming, was a joint venture of Kaneb Service, Inc., Koppers Company, Inc., Metropolitan Life Insurance Company, and Northwestern Mutual Life Insurance Company. It proposed to convert coal to 21,000 barrels per day of gasoline and other liquid petroleum products using KBW and Lurgi gasifiers, methanol synthesis, and the Mobil MTG catalytic process. Koppers was the developer of the KBW entrained flow gasifier being proposed. From the perspective of technology, the project represented relatively low risk since versions of all the technologies had either operated successfully before, or were, at least under construction in other applications. On the other hand, the gasification technologies were not especially promising if one were concerned with developing the low cost technology for the future in that they were not pressurized, nor did they contain any of the features that developers of new technology were exploring. A view within the Corporation dissenting from this conclusion noted that the KBW gasifier was particularly suited to be built eventually at large scale and could have corresponding economic efficiencies.

The Breckinridge Project was to be located in Breckinridge County, Kentucky. The plant would produce 49,500 barrels per day equivalent of a full complement of refinery feedstocks using the H-Coal direct liquefaction process. The sponsors were to include Ashland Synthetic Fuels, Inc. (a subsidiary of Ashland Oil) and Bechtel Petroleum, Inc. Several features of the project were of particular interest to the Corporation. The H-Coal technology represented one of the three direct liquefaction processes that had been developed through the pilot stage in recent years and that were ready to be built at full commercial scale. In addition it was one of the few projects to be sponsored by an oil company (albeit one of the smaller ones).

The First Colony Project to be located in Cresswell, North Carolina would have produced methanol from peat using the KBW gasification process followed by the ICI synthesis process to convert the synthetic gas into methanol. The plant's capacity would have been 1,900 barrels per day of oil equivalent. The project was sponsored by Peat Methanol Associates, made up of the Energy Transition Corporation, Koppers Company, and J. B. Sunderland. This project had a number of particularly interesting features. It would have employed the same KBW gasifier that was proposed for the Hampshire Project, discussed above, and would have had Koppers Company, the developer of the technology, as a sponsor.

The feedstock, however, would have been peat rather than western coal. The use of peat made economic sense for the project, but eventually posed some political problems for the Corporation. While the nation has significant amounts of peat, they pale in comparison with the amount of coal. And, while the Act called for projects with a diversity of feedstocks (presumably including peat), the Board was never enthusiastic about this resource base.

More importantly, the project's sponsors were more far-sighted than most proposers in managing the risk and keeping the amounts of equity and governmental assistance to "modest" amounts. The project was a single train and was markedly smaller than most other projects applying under the Initial Solicitation. One chief consequence of this design, which lacked most economies of scale, was that the unit production costs were higher and higher unit guaranteed prices were needed to make the project viable (but total project assistance was relatively low). Although this approach was adopted by more and more projects as time went on, the positive aspects of these features tended to be downgraded at the time.

An additional feature of the project was to provide early experience with the charged political environment that the Corporation increasingly had to endure. As luck would have it, one of the partners of the Energy Transition Corporation was William Casey, the director of the CIA. The *Washington Post* chose to dramatize a potential conflict of interest. The fact that Casey was inactive in the Energy Transition Corporation, and that it did not have a direct equity interest in the project, was seemingly irrelevant to the *Post*.

The Cool Water Project was to produce medium Btu gas with a Texaco gasifier, and the gas, in turn, was to be used to generate electricity within an integrated combined-cycle plant. The plant, already under construction in Daggett, California, was to have a capacity of 4,300 barrels of oil equivalent (about 117 megawatts of electricity). The project was sponsored by Texaco, Southern California Edison Company, the Electric Power Research Institute, Bechtel Power Corporation, General Electric Company, JCWP, and the Empire State Electric Energy Research Company. Given the technology, the stellar sponsorship, and the fact that all the necessary equity was at hand, this was a very attractive project for the Corporation. But there were still two key problems. First, the project was designed only for a seven-year demonstration, whereas Congress had only permitted the Corporation to fund commercial production facilities. Second, inasmuch as construction was already under way, how could the project demonstrate that government assistance was truly needed for the project to go forward (another requirement)? These proved to be thorny issues.

The Memphis Project, which was to be located in Memphis, Tennessee, would have produced medium Btu industrial fuel gas with a U-gas gasifier. This gasifier had a fluid-bed design as distinguished from the KBW and Texaco entrained-flow gasifiers proposed for the three projects discussed above. The plant capacity would have been 8,400 barrels per day of oil equivalent. It was sponsored by the Memphis Light, Gas and Water utility, which was seeking alternative sources of gas supply. The project had strong local support. Thanks to earlier financial support from DOE the project arguably had the most advanced engineering and the best cost estimates of any of the projects under consideration at that time.

The WyCoal Project was to be located in Douglas, Wyoming, near some of the vast western sub-bituminous coal fields. It too had benefited from financial assistance from the DOE to prepare more developed design and cost estimates. The project was sponsored by WyCoal Gas, Inc., a Panhandle Eastern Company. It would have produced synthetic natural gas from coal at an initial capacity of 13,500 barrels per day of oil equivalent. Its conversion technologies would have been virtually the same as those used in the Great Plains Project which had just negotiated an assistance agreement with the DOE under the interim program (P.L.96-126 also authorized DOE to negotiate early contracts until the SFC was operational)—i.e., a conversion process consisting of Lurgi fixed-bed gasification, followed by gas cleanup, shifting acid removal, and finally methane synthesis.

The North Alabama Project was designed to produce methanol in Murphy Hill, Alabama. The process was to consist of gasifying coal in a Koppers-Totzek entrained-bed gasifier, followed by shifting, acid gas removal, and methanol synthesis. Plant capacity was to be 12,000 barrels of oil equivalent. The sponsors were Santa Fe International, Kidder Peabody, and Air Products and Chemicals. The project's sponsors and antecedents brought several interesting issues to the Corporation's considerations. Most importantly, the project had been under development for some time under the auspices of the Tennessee Valley Authority. Before the passage of the Energy Security Act, TVA had received a substantial appropriation—amounting eventually to about $125 million—for the development of what would have been the nation's first commercial scale synthetic fuels facility producing liquids. It was no accident that the project was to be located in Alabama, given that the Chairman of the House Appropriations Committee was Tom Bevill from Alabama. As a consequence, the project was being thoroughly designed, but, because it had not yet expended a great deal of its appropriations, it was not as "mature" as say the Memphis Project.

Some of the issues engendered by these antecedents were as follows. First, pursuant to the Energy Security Act, the Corporation could not provide assistance to another government agency such as TVA, and, despite North Alabama's substantial political connections, Congress could hardly provide it separate funding after having just passed the Act. Consequently, for North Alabama to be successful in receiving assistance, it had to be restructured under a consortium of private sponsors, which accounts for the participation of the firms listed above. This sponsorship led to another seemingly trivial, but potentially troublesome problem. Around this time Kuwaiti interests had acquired Santa Fe International. Some opponents of the Corporation were quick to question the idea of providing assistance to an OPEC entity. On the other hand, some took a realistic line: if OPEC wanted to invest scarce capital in the U.S. to give us alternative fuels capability, how could we complain?

Second, TVA's development of the project was conservative in that the design emphasis was on building a first-of-a-kind facility likely to work and work well, while minimizing technological risks. This philosophy led to the selection of the Koppers-Totzek gasifier, of German design, which had been built abroad in multiple locations. Thus, in terms of technological diversity, it was not a very interesting project. Nonetheless, it was eligible to receive assistance.

Finally, Tom Bevill was not a man to lightly antagonize. On at least one occasion, Ed Noble found himself in Bevill's office defending the Corporation's treatment of the project.

The Tennessee Synfuels Project, to be located in Oak Ridge, would have produced gasoline from coal. The processes proposed included KBW gasification followed by the ICI methanol synthesis process and the Mobil MTG gasoline process. The plant's initial capacity was to be 7,691 barrels per day of oil equivalent—ultimately, to be expanded to 50,000 barrels per day. The project's sponsor was Tennessee Synfuels Associates, a joint venture of Koppers Synfuels Corporation and Citgo Synfuels, Inc. This was the third project being sponsored by Koppers to use their KBW gasifier. This project had also applied to the DOE for assistance as part of the interim program and had been one of the leading contenders for receiving assistance in that program.

The Paraho Project was to produce crude oil from shale by means of room and pillar mining followed by surface retorting in a Paraho retort. The project was to be located in Uintah County, Utah, having an initial capacity of 10,000 barrels per day that would be expanded to 30,000 barrels per day. The project was sponsored by the Paraho Development Corporation, which was a small R&D firm that had built and developed a pilot retort. It, however, was funded by contributions from an array of larger corporations in the energy business.

The Calsyn Project was to use the Dynacracking process to upgrade heavy oil and bitumen from tar sands as a feedstock for conventional petroleum refineries. The plant was to be located in Pittsburg, California with a scale of 5,100 barrels per day. The project sponsor was Calsyn, a joint venture of Tenneco Oil Company, CDC Oil and Gas, Ltd., Alberta Oil Sands Technology and Research Authority, and Dynalectron Corporation.

The eligibility of this project for assistance was a question that taxed the Corporation's legal staff. The Act did not make all heavy oils eligible for assistance because considerable amounts of heavy oil were already being produced commercially and did not require assistance. Accordingly, the Act defined eligible heavy oils as those that would not be produced absent government assistance and that entailed the use of pioneer technology. Calsyn would not produce the feedstock, but would simply upgrade heavy oil more economically than other available processes. The issue troubling the staff had to do with potential feedstocks: if such feedstocks were available to be upgraded, then they were being already produced commercially, and, by definition, were ineligible. (This was potentially a classic *Catch-22* situation.) But the Corporation's legal staff concluded that if a project's superior upgrading economics made certain heavy oils commercial to produce that would not have been commercial absent the technology, then the technology was eligible for assistance under the Act.

The CoaLiquid Project was already constructed and producing 1,000 barrels per day of a coal-oil mixture at Shelbyville, Kentucky. The project was sponsored by CoaLiquid, Inc., which was requesting price guarantee assistance until energy prices rose sufficiently for the project to become economically viable.

Although this group of eleven projects was significantly fewer than the number originally applying under the solicitation, the staff was encouraged to find an array of projects that covered the spectrum of attractive synthetic fuel resources and technologies. Moreover, it was apparent that some of the projects that did not pass this review would continue their development given the prospect of being considered in a later solicitation.

Several features of this group of projects deserve further comment. First, although the group includes some small-scale projects, many of them tended to be larger projects with tens of thousands of daily barrel capacity aiming at economies of scale. These projects clearly hoped that with a continuing escalation of energy prices, they might not need the price guarantee subsidy at all—i.e., they would be able to produce a product at a price commercially competitive. In such an event, a price guarantee would simply provide the safety net against the vicissitudes of the energy markets. It seemed to be particularly important for the large oil companies not to appear to be taking government subsidies.

Second, a pattern of project sponsors emerged that was to hold throughout the course of the solicitations. Except in instances in which a single large company was fully committed to a project (for example, Dow Chemical on the Dow Syngas Project and Union Oil on the Parachute Creek Project), the sponsorship usually consisted of a consortium including the technology developer, an architect-engineer, a feedstock supplier, and, perhaps, an energy end-user or marketer. Among the above eleven projects—notably Hampshire, Breckinridge, First Colony, Tennessee Synfuels, Memphis, North Alabama, and Calsyn--sponsors tended to have economic interests in the project that went beyond a simple return on equity in the project. Unfortunately, for the larger projects in particular, the amounts of equity required were so large that they exceeded the amounts that such players were willing to risk in light of just these interests. Thus, generally, the initial sponsorships were in a position of having still to seek out other potential investors for the project to have the minimum 25 percent of a project's cost stipulated by the Energy Security Act (for cases in which a loan guarantee was being sought). Obviously, if no loan guarantee was desired, then the sponsorship had to raise 100 percent of the financing.)

Third, one could see the effect of the Interim Program of Feasibility Studies and Cooperative Agreements managed by the Department of Energy. The fact that the Hampshire and WyCoal projects passed the maturity review was in large measure a result of the work performed under the earlier assistance program. The Memphis Project also benefited from prior government assistance. North Alabama as mentioned before received prior assistance through the appropriation process to TVA.

Strength Evaluations

Having passed the Maturity Review, these eleven proposals were subjected to a two-phased Strength Evaluation. The evaluation included judgments about a project's technological soundness, potential for replication, and management capability. It further included a comprehensive economic viability study, an evaluation of product marketability, and an analysis of potential project compliance with all socio-economic, health and safety, and environmental regulations.

The two-phase evaluation was designed to save weaker or less interesting projects the expense of preparing proposals containing all of the information that would be needed in a full review. Thus, the first phase of the evaluation was conducted on the basis on the material contained in the basic proposal (or, in the case of the Initial Solicitation, in the supplementary materials required by the Board). Those projects that appeared strong, or, at least, appeared as though they could marshal such evidence, were passed into the second phase of the Strength Evaluation, during which time they were obliged to submit more detailed materials to substantiate the initial conclusions.

The Board was determined to move expeditiously given the slow start of the program. Accordingly, following the January deadline for submittal of supplementary materials, the Maturity Review was completed on January 18th, and the first phase of the Strength Evaluation decisions were taken by the Board in its March meeting.

Only five of the projects passed the Phase I Strength tests: Breckinridge, Hampshire, First Colony, Memphis, and Calsyn.[75] The ones that were dropped were not necessarily weak. CoaLiquid and Cool Water were eliminated because the one was already in operation and the other under construction, which would make it hard to justify assisting them. The others tended to lack the necessary sponsorship in terms of management and equity. These attributes, of course, were amenable to being strengthened in time for subsequent solicitations, and a number of the projects did resubmit proposals.

Only the Hampshire and Breckinridge projects, however, were admitted into negotiations from the Phase II Strength Evaluation[76]. The three other projects were taken aback by this turn of events. They were by no means weak or unattractive projects. As we shall see later, two of them negotiated terms of an assistance contract under the very next solicitation. The problem in the Board's eyes was that they had not yet created a sponsorship group able to detail key elements of a management structure for building and operating the project, in part because significant chunks of equity were still absent. Some of the sponsors protested the decisions by observing that a critical mass of sponsorship already was in place and that in the private sector additional equity investors would normally appear as the terms of deals became clearer. The Corporation was not yet ready to accept this line of argument (though it became expedient to do so as early as the next solicitation). The Board did, however, directly admit the projects into consideration under the second solicitation, which had a deadline for proposals around this time.

At this time, the careful reader would expect to be reading about the next stage of project evaluation, which was to have been negotiations leading to assistance agreements. It was certainly the Board's intent that Hampshire and Breckinridge swiftly negotiate such agreements. The Corporation, however, received an unpleasant surprise and the first of many lessons. The sponsors of the two projects, despite having been selected from a sizeable competition, decided that they were, after all, not interested in winning. The SFC staff was nonplussed. As soon as the Board made the selections, the staff attempted to set up meetings with the project sponsors to begin the negotiations, but was put off with the vaguest of excuses until the actual picture emerged.

The project sponsors were not by any means being feckless. Times had changed in the year and a half since they had submitted their proposals. Despite their having met the Corporation's ongoing requests for materials to carry out the evaluations, the sponsors were increasingly reacting to the less favorable climate for synthetic fuels. The expense of staying in the solicitation up to this point was nominal and allowed the sponsors to preserve their options. Now, they were obliged to make vast commitments of equity to the projects if they wished to receive assistance. Both Hampshire and Breckinridge, as proposed, were to be large multi-billion dollar facilities. None of the major sponsors had an appetite for having to raise between a half a billion and a billion dollars.

Both projects withdrew from the Initial Solicitation. They halfheartedly reapplied under the next two solicitations in a restructured mode involving a downscaling and a significantly decreased need for equity—a pattern to be repeated during other solicitations.

Bechtel took the lead on Breckinridge and analyzed a series of options over time and canvassed all the large companies having a potential interest in such a project. They never did manage to bring it all together.

THE SECOND SOLICITATION

So, two years after its creation, the Corporation headed into the summer of 1982 with a failed solicitation and no projects having been awarded assistance. But the staggered structure of the new solicitation process had just resulted in the arrival of 35 proposals in response to the Corporation's June 1, 1982, deadline for submission under the Second Solicitation[77].

The Board directed the staff to accelerate the evaluation process from the five months needed under the Initial Solicitation for the Maturity and Strength Reviews. The Board hoped to complete these reviews under the Second Solicitation so that the projects could begin preliminary negotiations in about three months from the point at which the proposals were received. Thus, the staff turned to the 38 projects (i.e. the 35 new proposals plus the three transferred by the Board from the Initial Solicitation).

Of the 38 proposals, 14 represented new projects. Twenty-four represented projects that had already applied once, been determined not to meet maturity or strength criteria of the Initial Solicitation, continued some development (in many, but not all, cases), and wished to try again. This pattern pretty much matched the staff's expectations that underlay the design of the overall solicitation process.

The breakdown of technologies among the 35 projects were: nine for coal liquefaction projects, five for coal gasification projects, seven for oil shale projects, nine for tar sands projects, one for a coal-oil mixture project, and four for projects that proposed to produce fuels by other means.[78] (This latter category, of course, tended to include the non-serious contenders. While the "kook" level was low, a few such stuck with the process for a number of years.)

Seventeen of the proposals passed the tests of the Maturity Review and were subject to a full strength evaluation by the staff[79]. Some of the specific criteria for the reviews were modified over suc-

PROJECTS PASSING THE MATURITY REVIEW OF THE SECOND SOLICITATION

Project	Size/Technology	Sponsor(s)
TSA	7,691 barrels of oil equivalent (BOED) Coal Liquid	Tennessee Synfuels Associates, Koppers, & Cities Service
Beluga	54,000 BOED Coal Liquid	Cook Inlet Region, Placer Amex
Mapco	17,900 BPD Methanol	Mapco, Inc.
First Colony	4,800 BPD Methanol	Peat Methanol Associates, Energy Transition Corp., J.B. Sutherland, Koppers, & Transco
Memphis	8,600 BOED Coal Gas	Memphis LG&W; Foster Wheeler; & Mapco, Inc.
New England Energy Park	17,000 BOED Coal Gas	EG&G; Brooklyn Union Gas; & Eastern & Fuel Associates
Cool Water	4,300 BOED Coal Gas	Texaco; SCE; EPRI; Bechtel; GE;ESEERC; & JCWP
North Alabama	28,000 BOED Methanol	Kidder Peabody; Santa Fe Int.; Air Products; Raymond Int.; Houston NG; & Peabody Coal
Riverside	4,000 BOED Coal Gas	Philadelphia Gas Works & United Engineers & Construction
Coaliquid	1,000 BOED Coal-oil	Coaliquid, Inc.
Paraho-Ute	39,500 BOED Shale Oil	Paraho Development Corp.
Kensyntar	6,300 BOED Tar Sands	Pittston; Ward Douglas Co.; & KSA Resources
Sunnyside	35,000 BOED Tar Sands	Chevron & Great National
HOPKern River	3,600 BOED Tar Sands	Cornell Heavy Oil Process; Placid Oil; SEDCO; Holly Corp.; Encore Petroleum; Lyco; Ladd Petroleum; & Robert Glaze
Enpex Syntaro	10,000 BOED Tar Sands	ENPEX; TESORO Petroleum; Pickens Energy; C.E. Lummus; & Texas Tar Sands Ltd.
Santa Rosa	4,000 BOED Tar Sands	Solv-Ex Corp. & Foster Wheeler
Calsyn	6,050 BOED Heavy Oil	California Synfuels; Tenneco; AOSTRA; Dynalectron Corp.; & Canterra Energy Ltd.

Table VII-l:

cessive solicitations to reflect experience by the staff in their application, but, in essence, they embodied the tests described earlier in this chapter. These projects are shown in Table VII-l above.

These projects were then subjected to the full strength evaluation prescribed by the Solicitation, and within the accelerated schedule set out by the Board, the staff presented the findings by the early fall. The Board found only twelve of the projects to have met the Strength Criteria: First Colony, Memphis, Cool Water, North Alabama, Paraho-Ute, Calsyn, New England Energy Park, Kensyntar, HOPKern River, Enpex-Syntaro, Santa Rosa, and Sunnyside.[80] These projects represented the most serious contenders for assistance by the Corporation. In addition to the projects described earlier, these included other projects we will examine individually.

The New England Energy Park (NEEP) project sponsored by EG&G, Inc., Brooklyn Union Gas, Eastern Gas and Fuel Associates, and Westinghouse Electric Company. The project, to be located in Fall River, Massachusetts, would have produced 17,000 barrels of oil equivalent per day (BOED) of three products: electric power, methanol, and methane. The Westinghouse coal gasification process would be the first step common to producing any of the outputs. The Westinghouse gasifier was of interest in terms of technology diversity because it was the only fluid bed gasifier other than the IGT (on the Memphis Project) being proposed and no fluid bed gasifier had been built in the world at commercial scale. The multiple product approach was also intriguing. This design, while involving additional process steps, attempted to optimize the economics of the project in light of the seasonal Massachusetts' markets for energy. Thus, it would generate electricity throughout the year, methane (synthetic natural gas) during the winter heating season, and methanol during the rest of the year when the higher gas volumes were not required.

The Kensyntar Project sponsored by Pittston Petroleum and KSA Resources. The project, to be located in Edmonson County, Kentucky, was designed to produce 6,300 BOED of heavy oil from tar sands using an in situ wet combustion process. This was one of the first tar sands projects applying to the Corporation and one of three that proposed using an in situ process—involving indirect underground processes that generally used application of heat to reduce the viscosity of the oil and allow it to be brought to the surface through conventional pumping techniques.

The HOPKern River Project proposed by the HOPKern River Associates, which was a partnership of Cornell Heavy Oil Process Company, Placid Oil, SEDCO, Holly Corporation, Encore Petroleum, Lyco, Ladd Petroleum, and Robert Glaze. It would have been located in Kern County, California, and would have produced 3,600 barrels per day of 14 degree API crude oil from tar sands using a proprietary steam drive process. The process involved in situ techniques, but differed in approach from the Kensyntar process. This project would have drilled columns in the earth from which pipes could carry heat from steam generated on the surface into the tar sands deposits.

The Enpex Syntaro Project sponsored by the ENPEX Corporation, TESORO Petroleum, Pickens Energy, C.E. Lummus, and Texas Tar Sands Ltd. It would have produced and upgraded 10,000 barrels per day of tar sands using an in situ steam drive process. The upgrading would use the LC-Fining process, licensed by C.E. Lummus. The project site was in Maverick and Zavala Counties in Texas. These tar sands were particularly unattractive for normal commercial extraction because of their high gravity and sulfur content. On the other hand, if these processes were successful, it would permit a substantial resource to be commercially available.

 The Santa Rosa Project sponsored by the Solv-ex Corporation and Foster Wheeler Energy Corporation to produce 4,000 barrels per day of bitumen from tar sands. The project would have been located on a tar sand resource near Santa Rosa, New Mexico. The production would entail conventional mining techniques followed by a proprietary extraction process developed by the Solv-ex Corporation.

The Sunnyside Project proposed by Standard Oil of California and GNC Energy Corporation to produce 35,000 BOED of bitumen from tar sands. The project would have been located in Carbon County, Utah, on one of the nation's richer tar sands deposits. The bitumen would be produced through mining and solvent extraction of tar sands, and then upgraded to a syncrude using conventional refining processes. The Board particularly found this project interesting because of the sponsorship by Standard Oil, one of the few large oil companies proposing projects to the Corporation. They would clearly have the financial resources and expertise to carry out a successful project.

It was with a mood of optimism that the staff moved out of the *evaluation* mode into the *negotiation* mode with sponsors. Faced with a dozen projects that had passed the strength evaluation and a desire to make the first awards by the end of the year, this could have proved formidable. But as could be expected, some projects were less ready than others to get down to earnest negotiations.

The staff began with those that had close to a full subscription of equity by sponsors and a clear notion of what they needed in the way of financial assistance to make a viable project.

"Letter of Intent"

It quickly became apparent that the first contracts could not be rushed. For example, while First Colony was eager to proceed and had most of the elements in place, it still lacked some equity (equivalent to a minor partner), which presumably they could find in short order, once it was clear what the terms of assistance were likely to be. Santa Rosa needed Foster Wheeler, who would be the major investor, to take over the management reins. Also, the project needed to verify certain aspects of its pioneer technology in a pilot plant and prove out the resource on their New Mexico site before a binding commitment could be signed. Such work, however, would be costly, and neither of the Santa Rosa sponsors was willing to make that order of investment absent a clear declaration of intent by the Corporation to provide financial assistance if the factors being validated proved out satisfactorily.

The negotiating staff needed to find a device to breakthrough these hurdles in an expeditious manner. The device was a letter of intent. Literally, such a letter was an expression of the Chairman's intent—subject to the negotiation of acceptable definitive agreements, compliance with necessary legal requirements, and the absence of material change in the financial prospects of the project—to make a recommendation to the full Board to approve an agreement for financial assistance as outlined in an attached term sheet. The letter was not, however, legally binding on the Chairman, the Corporation, or the project sponsor.

The letter was used to provide a framework for the continued development of the project—namely the letter, including the term sheet, was to serve as a basis for the preparation and completion of proposed contract documentation and as a reference against which the sponsor may proceed to complete all of the activities necessary for closing a transaction under the assumptions that:

The final transaction still had to be approved by the appropriate decision-making bodies of both the sponsor group and the Corporation; and

No change occurred that would materially adversely affect the prospects of the project.

The activities necessary for closing a transaction might have included continuation of engineering design and making such commercial arrangements as equipment and service supply contracts and marketing agreements. The letter was particularly useful for project sponsors seeking to finalize equity agreements or partnership arrangements, or to secure debt financing not guaranteed by the Corporation's loan guarantee.

The criteria for issuing a letter of intent were:

The Corporation believed all issues had been sufficiently discussed, evaluated and agreed to, except under certain circumstances where further work (outlined as a condition to closing) could be reasonably identified, the outcome of which was not expected to change the prospect for the project.

All the terms making up a complete term sheet had been negotiated. The definition of those terms had been standardized and the language was set.

All the equity was in place or sufficient assurances existed that equity funds would be available to the extent necessary to ensure that total project capitalization would be achieved.

As a practical matter, the key elements of the letters were the financial terms (importantly including total obligational authority, amount of loan guarantee, level of guaranteed price support, duration of support, and so forth) and the conditions placed on the project sponsors.

The letters facilitated progress on all of the projects that were to become important to the program. In virtually all cases, the sponsors proceeded to meet the conditions (not always with successful results - i.e. see the discussion of the Santa Rosa Project later). The sponsors were not willing to make such investments, often entailing the expenditure of millions of dollars, without some tangible prospects of eventually getting assistance. Indeed, in those early days, sponsors were hardly willing to make the considerable investments required to complete contract documentation.

The use of the letters quickly came under the scrutiny of those congressional opponents of the program, who challenged the legal authority for using them, requested an investigation by the General Accounting Office, and sent several formal letters from Congress requesting a justification for their use. In the end, all challenges were successfully addressed.[81]

The central issue simply was whether a contractual mechanism not specifically cited in the Act could be used. The argument in favor of its use was the letters were not legally binding on either party and were merely a statement of the Chairman's intent. Indeed, a number of project sponsors chose not to go ahead with the projects after having received a letter of intent and having met the conditions. Moreover, after Board members changed as part of the second Noble Board (to be discussed later), the Board blithely disregarded several letters of intent and reopened negotiations to reconfigure the projects and renegotiate the amounts of assistance being considered.

In the process of developing the letters of intent, the staff had to deal with practical aspects of the Energy Security Act's authorities and stipulated use of obligation authority. First and foremost was the level of subsidy to be provided to the projects. Project sponsors, of course, wished to get the best rate of return on investment possible and to minimize any financial risks they were likely to encounter. The Corporation, on the other hand, had a responsibility to use the taxpayer dollar prudently—i.e., provide no more assistance than needed to allow the project to go forward with sufficient financial robustness to weather the more likely adverse occurrences common to new technologies.

The uncertainties regarding future events and the technologies in question lent the principal source of complexity to the negotiations. For example, if one believed that future rates of inflation would be high, then the contract would have to allow for more obligational authority than would otherwise be the case. Similarly, if one believed that the cost estimates for the project contained substantial uncertainties, and that technology was such that extensive time would be required before it achieved full capacity, then the amount of loan guarantee would have to be relatively high, and the level of price guarantee to pay off the loan and provide a reasonable rate of return would have to be relatively high, as well.

Obviously, one's view of the world and the project would determine how one calculated the amount of assistance to be required. While the negotiations could, to a large extent, be kept anchored in the available fact base and sound financial analysis, there could be no escaping the pervading need for judgments of all kinds.

In the attempt to minimize the number of issues to be dealt with, the sponsors and the Corporation each undertook separate analyses and, individually, made those assumptions regarding extrinsic factors that seemed reasonable: for example, likely levels of future inflation, interest rates, oil prices and so forth. Then discussions could focus on ways to bring differences in estimates of needed obligational authority closer together.

While some issues were treated in an adversarial fashion, there was much common ground allowing for tradeoffs. After all, both parties wanted the project to be successful and have a normal commercial lifetime, so that negotiations were hardly a zero-sum game. While the Corporation did not want to provide an unwarranted amount of assistance, it recognized that assistance needed to provide a competitive rate of return. In terms of the tradeoffs, for example, if the sponsors demanded accommodation for a high rate of inflation, then the Corporation would require an increased amount of profit-sharing on the assumption that the price of oil would rise commensurate to the inflation rate in the future. Similarly, while the sponsors desired to protect themselves against the possibility of large cost overruns, they did not want to pad the cost projections, and the amount of guaranteed loan, if there was to be one, because the larger the estimated cost, the greater the amount of equity the sponsors would have to raise and pledge to the project. Also, the larger the commitment fees to the bank for the loan would be.

The Corporation staff had to gain a mastery of the nuances of the assistance authority that they could use. Some of the questions that emerged early in the negotiations included:

How could the Corporation avoid losing obligational authority if a project receiving a commitment of assistance failed to go forward and use the authority committed? The Act prohibited reusing assistance.

How could the obligational authority be applied with most effectiveness? The legislation provided "no-year" money, which effectively diminished with inflation and the passage of time. The more assistance that could be provided up front in a project's life, the less would be required overall from the government to achieve a specified level of return. Given the time value of money, the most effective assistance would have been a direct up front grant to a project. The Act, however, included no such authority. For most of the projects, the assistance mechanisms considered were price guarantees and purchase agreements—both of which did not provide any subsidy until the project actually produced output, and then, only in proportion to the output of the project over time. The Act, on the other hand, did not preclude some front-end loading of the guarantee payments. For example, some of the negotiated contracts provided that the level of price guarantee would be higher earlier in the projects life than later.

How did one have to account for obligational authority when both price and loan guarantees were involved? Ultimately, the legal staff determined that loan guarantees could be "rolled over" into price guarantees. For example, if a project were awarded a total amount of obligational authority of $100, initially composed of $60 in loan guarantees, and $40 in price guarantees, then as the project paid off the loan, then those monies would become available for further price guarantee payments. In effect, the entire $100 would have become available as subsidy to the project provided that it never defaulted on the loan. The concept at work was that the Act controlled the amount of government money that was at risk at any time for the project in calculating the amount of obligational authority being committed. As a loan was paid down, the amount at risk for the government diminished as well. Thus, they could then become available for price guarantee payments.

The Board authorized the first three letters of intent on December 2, 1982, for First Colony, Santa Rosa, and Calsyn.

First Colony

The letter of intent between the Corporation and the sponsors of the Peat Methanol Associates Project (i.e. the First Colony Farms Project) was signed on December 13, 1982.[82]

The signatories were the Chairman of the Corporation, C.R. Pullin, chairman and chief executive officer of the Koppers Company, Inc., and W.J. Bowen, chairman and chief executive officer of Transco Energy Company. While the Energy Transition Corporation and J.B. Sunderland would continue in the sponsorship, they would not be responsible for much or any of the equity to be raised, and, hence, were not signatories at the letter of intent stage of negotiations. In addition, there was to be at least one other partner. The two signatories committed only to take at least two-thirds of the project's equity, and the additional partner(s) would contribute the remaining equity. The need to search for the additional equity was the major shortfall in the ability to proceed to develop the final assistance commitment. The letter specified that the general contractor for the project would be a company jointly owned by Koppers and Davy McKee, or another company satisfactory to PMA and the SFC. In addition, the letter specified that PMA would enter into a peat-harvesting contract with Nello Teer, a wholly owned subsidiary of Koppers, or another company acceptable to both PMA and the SFC.

The objective of the sponsorship was to construct and operate a 60 million gallon per year peat-to-methanol facility near Creswell, North Carolina. The project was estimated to have a total cost (including contingencies) of $576 million and was scheduled to be mechanically complete by December 31, 1985. The key financial terms of the proposed assistance were predicated on these and other key assumptions, such as world energy prices. The importance of these assumptions was to become all too clear by the time that the project had met all the conditions to the letter of intent almost one year later.

Form and Amount of the Proposed Assistance

Maximum SFC Obligation: The aggregate amount of assistance was $465 million.

Amount of Loan Guarantee: $341 million of principal and $25 million of interest.

Amount of Price Guarantee: The initial amount of price guarantees available was to be $99 million. The loan guarantee was convertible to additional price guarantees as it was paid off, providing that the aggregate amount of SFC obligational authority committed at any time never exceeded $465 million.

Form of Price Guarantee: The Support Price was to be calculated monthly, beginning at $1.05 per gallon of methanol as of January 1, 1983, and adjusted in accordance with appropriate indices plus two percent per annum. At the SFC's option, in lieu of the foregoing, the Support Price could have been $1.23 per gallon adjusted by appropriate standard indices only.

Duration of Price Guarantee: During the first fifteen years after initial production from the plant, for each gallon of methanol produced and sold, the SFC would pay an amount, if any, by which the Support Price exceeded the Market Price. This Price Guarantee had pursuant to the requirement of the Act to be reviewed and the subject of possible renegotiation within ten years after initial production.[83]

Revenue Sharing with the Government: The project would have been obligated to share revenues with the government for seventeen years after initial production whenever the Market Price exceeded the Support Price. (Only be twelve years in the event that the price guarantees were not negotiated beyond the initial ten-year period.)[84]

The letter of intent also contained other refinements– e.g., the establishment of a Debt Protection Fund regarding the guaranteed loan, and specification of conditions under which the revenue-sharing payments would be deferred, or would go to prepay the guaranteed loan.

This draft agreement and the one with Santa Rosa (to be discussed next) were the only ones to attempt to have a project raise non-guaranteed debt as part of the financing. The concept was one of trying to minimize the government's exposure by finding lenders that would be willing to lend to the project on the basis of pledging the project's assets as security. The amount of non-guaranteed debt according to the term sheet would have been $63 million. After six months of effort, the project sponsors were only able to get commitments for a relatively insignificant amount of unguaranteed debt. The concept just was not practical for such pioneer plants.

Santa Rosa

On January 17, 1983, the chairman of the Corporation and the partners sponsoring the Santa Rosa Project (the Foster Wheeler Corporation and the Solv-Ex Corporation) signed a letter of intent[85]. The letter contained terms pursuant to which the SFC would provide financial assistance to the partners in connection with their construction and operation of a project near Santa Rosa, New Mexico, to produce bitumen from tar sands.

The project was to mine tar sands using conventional mining techniques and to extract bitumen there from using a new, proprietary extraction process developed by Solv-Ex. The tar sand deposit was in a tract located east of the Los Esteros Dam and had estimated recoverable reserves of at least 19.2 million barrels of bitumen.

The facility was expected to operate at least 300 stream days per year when it reached full capacity of commercial production, each stream day processing approximately 13,000 tons of tar sands to produce approximately 4,000 barrels of bitumen. That bitumen was expected to be sold to refineries within a 1500-mile radius of the project. The project was to be mechanically completed within 24 months after execution of the Agreement, and to be in commercial production within one year thereafter. The definition for commercial production was having produced an aggregate of at least 168,000 barrels of product during a period of 60 consecutive days.

The letter specified a number of conditions to be met by the sponsors before an Agreement would be consummated. These will be generally discussed below, but one was key to the events determining the fate of the project. Namely, Solv-Ex was to proceed with all reasonable speed to complete construction of, and to operate on Santa Rosa tar sands ore for at least 90 days (including a run of at least 21 consecutive days) in a pilot plant in order for Foster Wheeler and Solv-Ex to confirm the technical and economic feasibility of the Project. While Solv-Ex had demonstrated the process at a small scale using batch processes, it had still to show that the same results could be achieved reliably with larger, integrated processes using "run-of-the-mine" ore. The pilot plant construction was already well along, and this condition was anticipated only to be a prudent, confirmatory step, which would not present undue problems.

Form and Amount of Proposed Assistance

Maximum SFC Obligation: $42.6 million.
Amount of Loan Guarantee: $20 million of principal and $1.6 million of interest.[86]
Amount of Price Guarantee: $21 million. There would be a provision for conversion to additional price guarantee authority of up to $20 million of any unused loan guarantee principal commitment and loan guarantee principal made available upon repayment of guaranteed debt. The aggregate price guarantee commitments could not exceed $41 million.
Form of Price Guarantee: The Guarantee Price was to be the sum of $25 per barrel plus transportation charges not to exceed $2.50 per barrel (as of the fourth quarter of 1982 and adjusted quarterly by the GNP deflator). But in no event could the price guarantee payment for any barrel of product exceed $10 (adjusted for inflation).
Duration of Price Guarantee: The earlier of 8 years or 6 million barrels of aggregate product.
Revenue Sharing with the Government: The partnership would pay to the SFC an amount equal to 25 percent of its net cash flow after debt amortization and allowance for taxes for a period of eight years after the commencement of price differential payments. For the subsequent eight years, the partnership would pay the SFC an amount equal to 50 percent of its available cash flow with caps.[87]

The letter contained other provisions to protect the interests of the Government: e.g., the partnership would have established a Debt Protection Fund and there could be no distributions to the partners until all Debt Protection Fund deposits, Mandatory Prepayments, and Cash Flow Payments had been made.

Conditions to be Met Prior to Consummating an Agreement

Before the Corporation was prepared to enter into a final binding assistance commitment, the parents and the partnership had to comply with ten conditions, generally dealing with construction contracts, equity at risk, financing arrangements, and permitting. One, however, was to have unforeseen problems: i.e., the partners or their affiliates were required to have operated an integrated pilot plant for at least 90 days (including a run of at least 21 consecutive days) and confirm the technical and financial feasibility of the project.

These conditions were primarily for the protection of both Foster Wheeler, who would be putting up most of the equity, and the SFC, which would be guaranteeing the debt. If all the conditions could be met, the prospects for a successful project were excellent. And, at this time, there were no reasons to believe that the conditions would not be met. Nonetheless, significant time and investment would be needed to meet some of them—notably the extended run of the pilot plant. None of the partners were willing to make that kind of investment unless there was a high likelihood of the Corporation providing assistance. Thus, once again the utility of the letter of intent was shown.

In the event, the key condition was not met. In starting a mine to feed the pilot plant, the project determined that the tar sand resource was not as rich (by a couple of percentage points) and uniform as they had been led to believe. A couple of resource assessments available to the sponsors indicated a higher richness, for example an independent assessment done by the Laramie Technology Center for the SFC confirmed the methodology of these earlier assessments. However, the opening of the mine disclosed that the resource had bitumen distributed more sporadically than had been believed, which would result in crucially higher mining costs. As a consequence, the project did not go forward. But it did lead to a congressional hearing. Since this was to become part of a general attack by a congressional committee on the Corporation, the specifics will be left to Chapter 12 below.

Calsyn

On May 20, 1983, representatives of the U.S. Synthetic Fuels Corporation and parties proposing to form a joint venture—the Alberta Oil Sands Technology and Research Authority, Dynalectron Corporation, the Ralph M. Parsons Company, and Tenneco Oil Company—signed a letter of intent for the SFC to provide assistance to the joint venture[88]. The joint venture was to construct and operate a tar sands and eligible heavy oil upgrading project in Pittsburg, California.

The project was to be a reactor using the Dynacracking Process Technology to upgrade tar sand and heavy oil feedstock to produce fuel gas product, naphtha product, distillate product, sulfur, and other byproducts. The facility was expected to operate 330 stream days per year and process each stream day between 5,100 and 7,000 barrels of eligible feedstock depending on the mixture of feedstock used.

The most unique feature of the project vis-à-vis other projects considered by the Corporation is that it was essentially an upgrading facility. It relied on others to provide the feedstock. As mentioned earlier, this feature required the SFC to affirm that the feedstock was "eligible."

The somewhat metaphysical reasoning to make this determination began with the fact that substantial amounts of heavy oils had been found that were not being produced because existing upgrading processes could not pay producers enough to justify developing a field and still themselves make a profit selling the output in a competitive market consisting of other refiners and marketers. A new upgrading process, such as Calsyn's, that was particularly suited to upgrading heavy oils into premium products would mean that an upgrader could pay more for the heavy oil input (than other refiners could) and still be profitable. In that event, then prices offered for heavy oils would rise and more such reservoirs would be developed—thus increasing the production of synthetic fuels. While receiving assistance, such a project would only be permitted to use eligible feedstocks. The letter of intent identified several specific heavy oil reservoirs in California that were determined to contain eligible feedstocks.

Form and Amount of Proposed Assistance

Maximum SFC Obligation: $49.8 million.
Amount of Loan Guarantee: $49.8 million.
Fees to the Government: An administrative fee of $249,000 upon execution of the Loan Guarantee Agreement, and a guarantee fee of one-half of one percent per annum on the average principal amount of guaranteed debt outstanding each year.

Calsyn was the only project with which the SFC negotiated a potential assistance agreement that would consist of only a loan guarantee. In effect, the assistance would be predicated on the concept that commercial financing would not be available for this type of first-of-a-kind technology without guarantees external to the project. If financing could be made available, and if the upgrading technology worked as designed, then the project could be made profitable without further subsidy (such as price guarantees). The amount of loan guarantee was calculated on the basis of the project having eligible costs amounting to $71 million. The amount of the loan to be guaranteed by the Corporation was to be $49.8 million. The remainder of the financing would be the responsibility of the partners. And, of the latter amount, the letter recognized $2.8 million of sunk costs as being eligible.

The letter of intent contained a number of provisions related to when participation in the project could be terminated and to requirements affecting the financial operations of the project vis-à-vis the guaranteed loan similar to negotiated terms with other projects—notably with regard to requirements for a Debt Protection Fund, for mandatory prepayments, and for restricting payments to the joint venturers from the project. The letter also contained a number of conditions that had to be met by the joint venturers before the Corporation would sign a final assistance agreement, generally relating to contracts, leases and eligibility of feed stocks.

All three projects proceeded diligently to comply with the conditions governing their respective letters of intent so as to be able to consummate the agreements by which the government would provide assistance for the construction and operation of the synthetic fuels facilities.

Alas, it was not to be. Peat Methanol Associates took the better part of a year to obtain the remaining required equity in the project as intended by the letter. Unfortunately, during that period methanol prices weakened significantly, leading the sponsors to request the Board to increase the amount of obligational authority that was to be provided in the assistance agreement.

The Board concluded that it did not choose to approve a richer agreement than that contemplated by the letter of intent. Its concern was that the project, in effect, would get to negotiate twice with the Corporation, figuratively, they would get two bites of the apple. The concern that this might occur underlay the Board's original reluctance to negotiate terms of an agreement before all the equity was in place on the sponsor's side of the table.

Instead, the Board suggested that the project reapply under the Third Solicitation and, if it was still competitive, then higher amounts of assistance could be considered. For better or worse, the project sponsors chose not to do so. They felt that they had already gone to considerable lengths to assemble equity and to create the basis for a viable project. They may also have sensed that the Board had never been enthusiastic about the peat resource, which was relatively small compared to the nation's coal and oil shale reserves.

The Santa Rosa Project succumbed when it determined that the resource on the New Mexico site was not quite rich enough to sustain an economically viable project. The sponsors went on to obtain another site with demonstrably richer resources in Utah and applied again to the Corporation under a later solicitation.

It was never quite clear what happened to the Calsyn Project. No insuperable hurdles were ever brought to the Corporation's attention.

To whom and when would the Corporation make an award of

assistance? It would be to Cool Water, a fourth project from the Second Solicitation to receive a letter of intent, albeit months after the prior three projects had received theirs.

THE COOL WATER FINANCIAL ASSISTANCE AGREEMENT

The Board had rejected the initial proposal from the Cool Water sponsors in the Initial Solicitation. This rejection was not the result of the project being perceived as weak, or undesirable, or immature. To the contrary, the Texaco gasifier was a potentially attractive technology with extensive development behind it. The project had an impressive array of sponsors, who had already arranged the necessary financing. Moreover, it would not need a large amount of assistance from the Corporation. The problems with providing the project assistance were perceived by the Board as twofold. First, since the sponsors had already arranged the financing, and construction was underway, how could it be argued that government assistance was essential for the project to be built and operated. A finding along these lines would be required under the terms of the Energy Security Act. Second, the project, as configured by the sponsors, was to be a five-year demonstration. The Act only authorized assistance for commercial production facilities (which would presumably have a normal lifetime closer to, say, twenty years). Thus, the Board concluded that the project would not be eligible for assistance from the Corporation.

The sponsors took these perceived problems to heart, arranged to reconfigure the project, and submitted a new proposal under the Second Solicitation. The sponsors undertook to demonstrate that although construction had begun, they were prepared to abandon the project absent government assistance. This demonstration rested on the significant changes that had occurred in the projections of energy prices since the project had been launched. While around 1980, most prognosticators anticipated a rising trajectory of oil prices, by 1982, it was evident that prices had fallen back a bit instead and seemed to be headed for a long plateau. These changes undermined aspects of the sponsors' financing plan. Although all the capital had been raised for completing construction of the project, the funding for operating shortfalls was all to come from Southern

California Edison. If energy prices continued to rise, Edison's subsidies to the project would have been modest. Without a continuing rise in prices, however, Edison's exposure could well have been a couple of hundred million dollars. Edison stated that it could not provide such levels of subsidy and, indeed, would back out of the project unless the government was willing to provide assistance to the project that would, in effect, cap Edison's exposure. The Board eventually found this argument persuasive.

In the course of negotiations between the sponsors and the Corporation under the Second Solicitation, a number of important changes were made to the project to convert it from a "Demonstration" to a "Commercial" configuration. A second gasifier was added to the project to increase its ability to stay on line and continue production (and revenue generation) should one or the other of the gasifiers experience problems that prevented it from operating. The gasifiers were not equivalent. For instance, the main gasifier had extensive heat recovery equipment that promised high efficiencies while the added gasifier was to be a "quench" gasifier lacking the extra equipment, but still capable of operating close to the design output of the main gasifier.

During negotiations, the financial structure of the project was modified as well. With price guarantee subsidies from the Corporation, Edison's operating exposure was to be capped at $30 million. The subsidies were to run for a five-year period, at which time virtually all the project's debt would be paid down. Then, Edison agreed to acquire and operate the facility for an additional 15 years on the assumption that electricity rates would be sufficient to cover just the operating costs of the facility. However, if this assumption proved not to be the case, Edison would not be obliged to acquire the facility and operate it at a loss. Nonetheless, all financial projections suggested that the project should enjoy a full 20-year life if the technology performed as designed.

Therefore, the way was clear for the Corporation to provide assistance to the project pursuant to the constraints of the Act.

The Cool Water Program

The Cool Water Gasification Program's mission was to design, construct, test, and operate the 1,000-ton per day coal gasifier (with heat recovery) integrated with a 117-megawatt (gross) combined-cycle electrical generating unit. The sponsoring consortium included Texaco Inc., Southern California Edison, Bechtel Power Corporation, the Electric Power Research Institute (EPRI), General Electric, and the Japan Cool Water Program Partnership. The Empire State Electric Energy Research Corporation and the Sohio Alternate Energy Development Company were also project contributors.

The plant was located in the Mojave Desert near Barstow, California, next to Edison's existing Cool Water generating station. The plant was a commercial scale facility designed to process a variety of coals, operating during the first years predominately on high-grade bituminous coal from Southern Utah Fuel Co. mines near Salina, Utah. A variety of other coals (notably high sulfur Eastern bituminous coals such as Illinois #6) were successfully used in test programs throughout the early years of operation. Water for the plant was obtained from wells at the generating station. An adjacent air-separation plant owned and operated by AIRCO supplied oxygen for the coal gasification process.

The plant produced a medium-Btu fuel gas with a heating value of approximately 265 Btu per standard cubic foot (SCF). At design capacity of 1,000 tons per day of coal feed (dry basis), total syngas production was approximately 70 million SCF per day. The plant also produced 10 million pounds per day of high-pressure superheated steam. Both the syngas and the steam are used in a combined-cycle unit to generate about 111 megawatts (net) of electric power, which was to be sold and transmitted through Edison's existing power distribution system. The amount of electricity produced by the Cool Water project was equivalent to that obtained from an oil-fired unit consuming 4,300 barrels of oil per day and can satisfy the daily electric needs of 50,000 residential customers.

The technology used in the project offers a number of improvements over other gasification technologies that had been commercially available:

Feedstock flexibility: the gasifier is capable of processing all ranks of coal, but its performance is more attractive with higher ranked coals;

Pressure flexibility: the product gas can be produced at the consuming pressure, making additional compression (and cost) unnecessary;

Rapid process response: the system can respond to abrupt changes in fuel delivery requirements;

High environmental acceptability; and

Product versatility: the product gas composition makes it useful as a chemical feedstock as well as a fuel.

Aside from its relative attractiveness relative to older gasifiers, the integrated coal gasification combined-cycle technology (IGCC) used at the Cool Water facility was to be one of the most efficient and environmentally sound methods of fossil fuel based power generation in the United States. The plant's operation demonstrated that the gasifier and the combined-cycle unit can be integrated efficiently and that power can be dispatched reliably.

The Price Guarantee Commitment

The Corporation concluded a financial assistance agreement with the signing of the Price Guarantee Commitment (PGC) on July 28, 1983,[89] with the following features:

Maximum SFC Obligation: $120 million.

Amount of Price Guarantee: $120 million provided in the form of price guarantee payments on the medium-Btu syngas produced (eligible product). The plant converts the gas to electric power, which was sold to Edison and distributed through the Edison system. Because the Act defines gas as an eligible product, but not electricity, the payment system converted the electricity sold back into quantities of gas, which become the quantitative basis for calculating the price differential payments.

Form of Price Guarantee: The price guarantee payments were calculated on the basis of a two-tier set of guaranteed price for eligible product—i.e., $12.50/million Btu (plus escalation for inflation) for the first 9 trillion Btus; and $9.75/million Btus (plus escalation) for the next 11 trillion Btus.

Duration of Price Guarantee: The Corporation's financial assistance was to stop when 20 trillion Btu of eligible product had been produced or when $120 million in price differential payments had been made, whichever occurred first.[90]

Revenue Sharing with the Government: Net program revenues were to be shared first to the SFC an amount in
excess of $100 million up to a maximum of $20 million,
and second between the SFC (12.5 percent) and the participants.

When the reader has become more familiar with the amounts
of obligational authority negotiated with other project sponsors, it
will become apparent that the amount of assistance provided to
Cool Water was relatively low. The reason for this outcome was
twofold. First, the utility industry through EPRI was itself providing a subsidy, and a number of the participants were also subsidizing the project in order to further the development of technologies
in which each hoped to receive future financial returns.

Second, the project was already about 60 percent completed at
the time that the financial assistance was awarded. The investment
would be effectively lost if the plant were not completed. Consequently, the Corporation adopted a negotiating posture that it was
dealing with a "workout" situation where it would help cap the
future financial risk of the participants (principally Edison), but it
would not attempt to help ensure a rate of return. In other project
negotiations, the Corporation, of necessity, had to base the amount
of assistance on an amount assuming no subsidy by the sponsors,
and providing a rate of return commensurate with the risk being
assumed by those sponsors in developing a first-of-a-kind technology. Thus, the amount of assistance provided to the Cool Water
Program by the Corporation should not be viewed as a standard for
measuring the assistance offered to other projects by the Corporation.

Contract Monitoring

The administration of the PGC was straight forward and in keeping
with the hopes of the framers of the Energy Security Act: the project
sponsors retained full authority and responsibility for the sound
commercial design of the facility, and for cost-effective construction
and operation. With a fixed guaranteed price level, if the management proved weak, then the participants would have to cover the

costs involved. By the same token, the Corporation's views were not allowed to confuse project management, extend decision-making, or otherwise undermine the project's commercial orientation.

The price guarantee mechanism established by the PGC proved to be easy in implementation: (1) the market price was set by rates approved by the California Public Utility Commission, with those rates fixed throughout the five years of Stage I operation, and (2) each month the Program would measure how much gas it had produced and would bill the Government for the difference between the guaranteed price (e.g. $12.50/million Btu for the first 9 trillion Btu) and the approved rates multiplied by the quantity produced. The monitoring tasks were equally simple: the SFC team would assess the instrumentation and measuring procedures, would check the price indices to adjust the guarantee price, and would arrange to have audits performed of all documentation.

There were, of course, other requirements of the contract to be followed, notably the program had to develop and implement an Environmental Monitoring Plan. To ensure that all areas of environmental concern were identified, the Corporation required supplemental monitoring (i.e. monitoring not required by permits) of project emissions, and worker health monitoring as part of the Plan. For example, part of these requirements included a worker registry that tracked employee health for the life of the project. The data collected as part of supplemental monitoring was reported quarterly and annually to an inter-agency Monitoring Review Committee, and is available to the public, and, of course, to any groups interested in replicating the facility.

OVERALL SFC PROGRAM STATUS

As an institution, the Corporation had come through its initial shakedown period with a dedicated Board, a capable focused staff, and renewed momentum tempered by the changing realities of energy markets. Specifically, the staff had shown that it could evaluate large numbers of complex projects quickly and insightfully as summarized in Table VII-2.

In tandem, the Board had exhibited a sense of common purpose and an ability to act decisively with regard to projects under

OUTCOME OF THE INITIAL AND SECOND SOLICITATIONS					
	Projects Submitting Proposals	Projects Meeting Maturity Criteria	Projects Passing the Strength Revies	Letters of Intent Authorized	Projects Awarded Assistance
Initial Solicitation	68	11	5	0	0
Second General Solicitation	38	17	11	4	1

Table VII-2

consideration. Anticipating an intensified pace following the issuance of additional solicitations (see later chapters), in August 1982 the Board established the position of executive vice president and appointed Jimmy Bowden to it. He had extensive background in energy projects generally, and coal synthetic projects, in particular, from his prior positions with Conoco.

As activities on the first two solicitations wound down, the Corporation had assembled the makings of a program represented by an array of mature, potentially strong projects that reflected a satisfying diversity of technology and resource: coal liquid, coal gas, coal-oil mixtures, peat, oil shale, and tar sands projects from which to make awards. And two projects were under construction: the Parachute Creek oil shale and Cool Water coal gasification plants.

There were, however, clouds on the horizon. 1982 was a year of economic recession, during which U.S. imports of oil fell to about half that previous, and energy prices, particularly of refined products, fell as well. In addition to the changed economic and energy outlook, federal tax policy changes had a discouraging impact on the investment decisions of some project sponsors. With the passage of the Tax Equity and Fiscal Responsibility Act of 1982, synfuels investors faced a reduction in the Accelerated Cost Recovery System (ACRS) basis for tax credits taken, a repeal of 1985/6 ACRS schedule changes, a capitalization of certain construction period interest and property taxes, and no extension of the Energy Tax Credit beyond 1982 in most cases. The net effect of these changes was to make prospective sponsors have to invest more capital in a given project than would otherwise have been the case. And, as noted earlier, raising the capital for projects was already appearing as the greatest problem facing the Corporation in terms of its ability to expeditiously conclude assistance agreements.

The effects were already seen among project sponsors. As noted, with the changing energy environment, the Breckinridge and Hampshire sponsors increasingly appeared to be reluctant brides, while Exxon had announced its decision to suspend work on the Colony shale oil project and the award to Tosco had been terminated.

All this notwithstanding, the Corporation had made significant progress in lining up a beginning slate of attractive diverse projects and was about to pick up the pace with the issuance of additional solicitations of proposals.

VIII

Casting a Wider Net

During the latter part of 1982, with experience from the Initial and Second General Solicitations in hand, the Board decided to explore some other solicitation options. Experience to date seemed to indicate that the solicitation process was not attracting the full range of technologies that would be necessary to meet the Energy Security Act's diversity goals. Notably, the process had attracted few mature oil shale proposals. In addition, relatively few of the large energy firms with "deep pockets" seemed willing to participate in the program.

There were three contributing factors that especially shaped the Board's view. First, some limited discussions with private firms indicated that major potential players had not participated in the ongoing solicitations because of their discomfort with the vagueness of procedure and outcome. Also, many of the large oil companies had longtime predilections of not participating in government programs of this type because of the likelihood of inevitable political rhetoric characterizing any assistance as subsidizing the "fat cats." Moreover, they were traditionally sensitive to the government intruding into the energy sector. They would have particularly disliked the Corporation undertaking a government-owned facility, for which the Corporation had a "last-resort" authority as discussed earlier.

Even putting those inhibitions aside, there appeared to be uneasiness that the general solicitation procedures were too subject to manipulation by the Board. Some of these companies, presumably, did not want to go to the effort and expense to prepare a proposal and to attract the publicity of applying to the Corporation, only to have the Board change the rules of the game along the way to everyone's frustration. This reasoning was unduly negative given that the Corporation had set out a transparent, good-faith procedure as part of the general solicitations that should have met everyone's needs. Nonetheless, some future events were to show that these reservations were not unfounded. For example, under the second Noble Board, as we shall see, some of the sponsors found that political circumstances forced negotiated agreements to be renegotiated. After a solicitation had closed, the Board began to pressure sponsors with regard to project size and the configuration of technology that they were willing to entertain. Some sponsors prepared multiple revisions to their proposals to react to the Board's guidance, and then still did not receive any assistance because of congressional reluctance.

Furthermore, members of the Corporation believed that it should try a more competitive approach that would minimize staff evaluation and negotiation time in arriving at an award of assistance. Staffers who had trained in economics wanted to see if the SFC could not simply specify the type of facility it was interested in, let qualified firms bid competitively for assistance, and then let the Board select the lowest bid as the best deal for the government.

Finally, such an approach might clear the way for the Corporation to work directly with interested firms to reach agreement on assistance for certain key projects outside the solicitation process. Congress had been quite clear in the Act that it wanted the Corporation to select projects competitively whenever possible. In Section 131(b)(4), however, the Act stipulated that where a formal solicitation for competitive bids fails to result in acceptable bids, then the Corporation could undertake direct negotiations with potential sponsors. So, in part, the Board was trying to create procedurally another option for supporting projects, should the general solicitation process fail to generate sufficient proposals.

Thus, the Corporation embarked upon targeted solicitations: three competitive low-bid solicitations were ensued in short order. The resulting experience, however, led to a significant recasting of the approach producing an additional five "modified targeted" solicitations.

TARGETED COMPETITIVE LOW-BID SOLICITATIONS

The specifications for the new solicitation approach were to structure a situation where potential sponsors could make a meaningful financial bid that could be evaluated competitively with other such bids. One difficulty was that potential sponsors would be interested in different scales of facilities, producing different product mixes, and requiring different forms of available assistance (i.e., price and/or loan guarantees). Thus, at a minimum, the solicitation would have to define a desired project such that sponsors could have a reasonable prospect of configuring a project and competitive bid that could be a winning submittal. If undefined parameters and uncertainties were too great, it would be unlikely that sponsors would devote the time and effort that would be required.

Specifically, the solicitation would have to give more definition of desired projects to potential bidders than had been the case in the general solicitations. Accordingly, they specified the resource category eligible for assistance, a minimum daily production capacity, a deadline for the initiation of project operations, a minimum lifetime operating period, and a maximum dollar limit on the amount bid.

There were ultimately three targeted low-bid solicitations: (1) Western oil shale, (2) Gulf Province lignite gasification, and (3) Eastern Province and Eastern Region of the Interior Province Coal Gasification.

Western Oil Shale

The first of the targeted solicitations was issued on January 20, 1983, aiming at the oil shale resource, which represented a particularly large resource suitable for synthetic fuels and which was sparsely represented by projects in the first general solicitations. It invited proposals for projects using western shale to produce a minimum level of 10,000 barrels of oil equivalent per day (BOED), with operations to begin before 1990, and committing to a facility that was to have a minimum operating life of 20 years. The solicitation stated that the Corporation could disqualify all competitive bids in excess of $1.6 billion.

Within this framework, the solicitation's evaluation and negotiation procedures were to be different from those of the general
solicitations. There were to be two stages. The sponsors were first
required to submit qualification proposals demonstrating their capability to finance, design, construct, and commercially operate the
production facilities: the qualification proposals were due March
15, 1983. The Corporation's staff would evaluate the proposals
against the criteria contained in the solicitation. While the evaluation techniques were similar to those of the earlier solicitations,
the data requirements were less onerous and the evaluations could
proceed more expeditiously.

In a second stage, sponsors that were designated as "qualified
bidders" were invited to submit a detailed technical proposal and a
competitive bid. One of the chief differentiating criteria among proposals was a demonstration of the equity committed to the project.
For the qualification proposal, sponsors had to demonstrate that a
minimum of 20 percent of the required equity had been committed.
At the time of the Technical Proposal, the minimum demonstrated
equity amount was 60 percent.

The solicitation specified the form for the competitive bid, which
was to accompany the Technical Proposal in a sealed envelope. In
effect the bid amounted to the maximum SFC liability (obligational
authority to be set aside in accordance with the procedures specified by the Energy Security Act) for either loan guarantees or price
guarantees. Where a bid included a combination of price and loan
guarantees, the bid was to be multiplied by a competitive param-
eter of 1.05. This latter feature was a device to give some advantage
to proposers that sought only one form of assistance on the basis
that the Act directed that dual forms of assistance be granted only
where one would be insufficient to achieve a financially viable project and to achieve the purposes of the Act.

In response to the targeted oil shale solicitation the Corporation
received a total of six Qualification Proposals, of which only four
were found to be qualified. None of the four came from new proposers that had not previously applied to the Corporation under
one of the general solicitations. Three of the projects were already
under consideration in the Third General Solicitation—Union Oil
Parachute Creek Shale Oil Program Phase II, White River, and Ca-

thedral Bluffs. The fourth project, Syntana, had been submitted under the Initial Solicitation.[91]

Of the four qualified proposers, only Union Oil submitted a Technical Proposal and a Bid. The Board found the bid to be nonresponsive. Union had proposed the same 40,000 BOPD facility that was under consideration in the Third Solicitation and which would require substantially more assistance than the $1.6 billion reservation bid that the Board had established for the targeted solicitation. Consequently, the Board terminated the targeted oil shale solicitation without making an award.

Gulf Province Lignite

In February 1983, the Board authorized the preparation of a targeted solicitation for Gulf Province lignite gasification projects. Gulf Province lignite represents a substantial resource in terms of its size, its suitability for use by a synthetic fuels industry, and its proximity to energy intensive industries in the lower Mississippi Valley and Gulf Coast region. No other major coal resources exist in that area to offset declining production of conventional oil and gas.

The targeted solicitation requested projects proposing the commercial production of medium-Btu gas, high-Btu gas, or methanol (or a combination thereof) by 1991. The minimum project production scale was to be determined on a Btu basis and was specified to be 10,000 barrels per day on crude oil equivalency (COE). As with the competitive solicitation for oil shale, projects were limited to competing for assistance in the form of a price guarantee, a loan guarantee, or a combination of the two. The Corporation again reserved the right to reject any competitive bid over $1.6 billion.

The Arkansas Lignite Conversion Project, sponsored by Arkansas Power & Light (a subsidiary of Middle South Utilities, Inc.), International Paper Company, and the Dow Chemical Company, was the sole respondent to the July 25, 1983, deadline set by the solicitation. The project, which would have been located in Hampton, Arkansas, proposed to gasify Gulf Coast lignite utilizing the Dow gasification process to produce 66.2 billion Btus per day of fuel gas (11,400 BOED).[92]

The Board found the project to be qualified and invited the

sponsors to submit a Technical Proposal and Bid. This the sponsors did, but the proposal and bid were found by the Board not to meet the requirements of the solicitation. The Board terminated this solicitation, like the prior oil shale solicitation.

But the Board did not want to abandon a promising project sponsored by substantial companies. Consequently, the Board pursued the provisions of the Act that permitted direct negotiations with sponsors to arrive at an assistance agreement—i.e., in the absence of acceptable proposals to a competitive solicitation, and given that the resource was considered essential to a complete diversified program mandated by the Act, Section 131(b)(4) permitted direct negotiations. Such findings were required to be reported to the Senate Committee on Energy and Natural Resources and the Speaker of the House of Representatives, and the Board did so.

The Corporation was not, however, restricted to the Arkansas Lignite Conversion project for negotiations. To assure that no reasonable proposal to develop Gulf Coast lignite was overlooked, the Board directed the staff to attempt to identify any other likely projects that could be considered competitively with the Arkansas project. No such projects were identified.

In the course of the direct negotiations with the sponsors of the Arkansas project, the sponsors proposed a significant increase in the amount of assistance they would require and proposed a decrease in project size. This placed the Corporation in a quandary. There was much they liked about the project and the sponsors, but being the only project in the game, they were pressing the assistance levels. More important, from a legal perspective, one could argue that the smaller facility with higher assistance was no longer in the bounds established by the solicitation as having no other interested parties. Perhaps, if others knew originally that the Board would have considered richer terms than that laid out in the solicitation, they too would have applied. There was no way to be sure.

The Board felt that it had no choice but to discontinue the direct negotiations. Thus, the second of the targeted solicitations also had no issue.

Eastern Coal Gasification

The third targeted low-bid solicitation, for Eastern Province and the Eastern Region of the Interior Province bituminous coal gasification projects, was issued in June 1983. That solicitation followed closely on the heels of the Gulf Province lignite solicitation because of the importance of Eastern coal. Eastern and Midwestern bituminous coals represent almost half of the domestic coal reserves and are located near water, transportation, and major industrial centers of the United States.

The solicitation was first limited to projects producing medium-Btu gas, high-Btu gas, methanol, or gasoline (or a combination thereof) and proposing commercial operation by July 1991. The solicitation was later amended to include ammonia and ammonia synthesis gas as eligible products. The solicitation initially stipulated that the minimum scale of projects was to be determined on a Btu basis and was specified to be 10,000 BOED. Minimum scales for the ammonia plants were set at 2,400 short tons of ammonia per day or 215 million standard cubic feet of ammonia synthesis gas per day. As with the first two targeted solicitations, sponsors were limited to competing for a price guarantee or a loan guarantee, or a combination therefore. The Corporation reserved the right to reject any competitive bid over $1.9 billion.

This targeted solicitation added one significant additional step to the proposal/evaluation process. It required potential proposers to respond to a pre-qualification step, which would provide the Corporation with an early indicator as to the interest that existed among potential sponsors. The step asked sponsors to affirm their intentions to submit a proposal and to provide evidence of the sponsor having a reasonable prospect of assembling the total financial resources that would be necessary to carry out the project (with Corporation assistance, of course). This step could have provided data for a finding to go to direct negotiations, should interest prove nil and should such a move seem advisable to the Board.

Nine projects responded to the pre-qualification step. Eight of these were found to have satisfied the requirement for demonstrating the capability for assembling financial resources for constructing and operating the facility being proposed. Only five of these, however, met the solicitation's December 1, 1983, deadline for submission of qualification proposals. (An affirmation of the sponsor's intent to submit such a proposal made as part of the pre-qualification submittal was not enforceable.) One of the five withdrew from the solicitation shortly thereafter.[93]

In the following January, the Board selected three of the four projects as qualified bidders under the terms of the solicitation:

The COGA-1 Project, which was to be located in Southern Illinois, would have gasified 6,600,000 tons per year of Illinois #6 coal utilizing the Texaco coal gasification process to produce 1.1 million tons per year of anhydrous ammonia. The sponsor group at the time identified Dean Witter Reynolds and CBI Industries, Inc. as equity participants.

The Sharon Steel Project, which would have been located at the Sharon Steel plant in Farrell, Pennsylvania, would have gasified Pittsburgh seam bituminous coal to produce 280 million standard cubic feet per day of medium-Btu gas utilizing the Kloeckner molten iron coal gasification process. Project participants included Sharon Steel Corporation, DWG Corporation, and Kloeckner Kohlegas GmBH.

The Louisiana Synthetic Fuels Project, which would have been located at Georgia-Pacific's plant in Plaquemine, Louisiana, would have gasified eastern coal to produce 58 billion Btus per day of medium-Btu gas utilizing the British Gas/Lurgi slagging gasification process. The sponsoring group consisted of Babcock Woodall-Duckham, Ltd., British Gas Corporation, and the BOC Group. Georgia-Pacific would have supplied the site, have operated the plant, and would have bought the product.

Only one of the above three, the Louisiana Fuels Group, submitted a Technical Proposal. As in the case of the other targeted solicitations, the Board found the bid to be unresponsive, and then, terminated the solicitation. All three projects continued development, however, in somewhat different guises, and reapplied under subsequent solicitations. Two of them might well have resulted in assisted projects.

THE FIVE MODIFIED TARGETED SOLICITATIONS

The targeted competitive solicitations clearly did not achieve the objectives that had been set for them. That is not to say that the solicitations did not attract potentially good projects. They did. But they did not result in any awards, although both the sponsors and the Corporation continued to be interested in enabling a number of the projects to come to realization.

In large measure, it was the solicitation process that was flawed. It was well defined and would have minimized the Corporation's ability to manipulate sponsors. But, as a practical matter, the process was simply too rigid for projects of this magnitude and degree of risk. Sponsors did not want to lock themselves in a specific "low bid" while its design was not complete and cost estimates continued to evolve. Other sponsors were not interested in building a facility as large as the minimum size specified by the solicitation.

There was, however, nothing dysfunctional in having a solicitation target an area—or type of project—that was of particular interest to the Corporation in completing a diversified program. The next five solicitations were all targeted to specific areas. They did not attempt a competitive low bid process and were predicated on following the wide-ranging negotiating process used by the General Solicitations.

Solicitation for Coal or Lignite Gasification Projects

The first of the modified targeted solicitations was a Solicitation for Coal or Lignite Gasification Projects, which was issued on January 5, 1984, arising from a request from the Great Plains Project for assistance from the Corporation. It arose because the Corporation was not empowered to negotiate assistance for a particular project outside the bounds of a competitive solicitation, unless, as discussed earlier, it could demonstrate that it had attempted to attract projects in a given area and had failed. So, before the Corporation could discuss assistance with Great Plains, it had to prepare a solicitation and attempt—with transparency and fairness—to attract other projects with which Great Plains would have to compete. Although honest attempts were made in this direction, an unbuilt project burdened with major cost and operational uncertainties could not really hope to compete on the basis of a low bid.

The Great Plains Project, as indicated earlier, had received assistance from the Department of Energy in the form of a loan guarantee pursuant to the authorities of the Non-nuclear Act and the legislation authorizing the Department to carry out an interim program until the Corporation was declared "operational." Loan guarantees, however, alone do not provide a project with a production subsidy to improve the economics. For this reason, none of the awards eventually made by the Corporation were loan-guarantee only. The downside risk associated with fluctuations in future energy prices was too great to ensure financially viable project operation over an extended time frame. For whatever reasons, the Great Plains sponsors and DOE originally believed that the special tariff that the project had won from the Federal Energy Regulatory Commission would make the project fully economic and that the sponsors simply needed a loan guarantee to raise the capital necessary to build and operate the project. This tariff guaranteed the project a higher rate for its syngas than was being paid to natural gas producers. The problem, it turned out, was that the special rate was good only until 1989. Thereafter the project would still have a beneficial rate, but it was tied to Number 2 fuel oil. With the decline in all energy prices, it became apparent to the sponsors that the project might not survive the new tariffs in 1989. If that were the case, the sponsors wished either to terminate their involvement in the project without sinking any additional capital into it, or to obtain additional subsidy from the Corporation.

The Board was initially unpersuaded by the arguments made by the Great Plains sponsors and was disinclined to issue a solicitation. Nonetheless, they had the staff analyze the financial data provided by the project sponsors to assess the arguments of unviability being put forward. Consequently, the Corporation's financial staff did extensive work using the Corporation's financial model, producing an ambiguous conclusion that the sponsors' assertions could not be disproved. In effect, the result hinged greatly on the assumptions one made regarding future energy prices as well as whether the sponsors would actually abandon hundreds of millions of dollars of investment (the plant was largely built) on the basis of energy projections five years in the future.

In any event, the Board chose to issue a Solicitation under which Great Plains could apply for assistance. Any potential project sponsor could request the Board pursuant to Section 127(c) of the Act to issue a general solicitation requesting proposals under which it could be competitively considered for assistance. While the Board did not inevitably honor such requests, it did do so on a few occasions when it appeared that a viable project might not otherwise have an opportunity to be considered. Thus, the Board believed that it would be consistent and fair if it issued a solicitation in response to Great Plains' request.

Consequently, the Solicitation for Coal or Lignite Gasification Projects sought projects producing a minimum of 10,000 barrels of crude oil equivalent by 1990 in the form of medium-Btu gas, high-Btu gas, methanol, gasoline or liquid petroleum gases. Only price guarantee assistance was offered. The solicitation stated that preference would be given to projects offering the earliest production at the lowest unit cost, and requiring the least financial assistance per unit of production and the least total amount of financial assistance overall.

Although some other potential applicants had been interviewed by Corporation staff and an effort had been made to broaden the boundaries of the Solicitation in terms of project minimum size and variety of eligible product, no project other than Great Plains applied under the Solicitation.[94]

The Great Plains Coal Gasification Project was sponsored by the Great Plains Gasification Associates, a North Dakota partnership comprised of Tenneco SNG, Inc., ANR Gasification Properties Co., Transco Coal Gas Co., MCN Coal Gasification Co., and Pacific Synthetic Fuels Company. Its design capacity was 137.5 MM standard cubic feet per day of high-Btu synthetic gas from 14,000 tons per day of North Dakota lignite using the Lurgi coal gasification process, and was to start up in July, 1984 (which it did).

After finding the project qualified under the solicitation and negotiating potential terms of assistance, the Board authorized issuing a letter of intent on April 26, 1984, with the following terms:

Form and Amount of Proposed Assistance

Maximum SFC Obligation: $790 million in price guarantees.

Form of Price Guarantee: For the first three years, $10.00 per million Btus of eligible product; for the duration of assistance, $7.50 per million Btus of eligible product, both adjusted monthly for inflation by the Producer Price Index, excluding food.

Duration of Price Guarantee: Ten years from the in-service date or until the maximum of $790 million had been reached.

Revenue Sharing with the Government: As long as DOE guaranteed debt was outstanding, 90 percent of after tax cash flow would prepay it. Thereafter, profit sharing payments of 70 percent of after tax cash flow would go to the SFC.

Estimated Net Cost: The estimated net cost of assistance, net of profit sharing and taxes over the 30 year life of the project, was 40 cents per barrel of oil equivalent in 1984 dollars.

As a condition of assistance, the sponsors had to invest an additional $100 million of equity and had to agree to apply all tax benefits and profits for the prepayment of DOE guaranteed debt for approximately three years after contract closing.

As luck would have it, April 26, 1984, the date of authorizing the letter of intent, also turned out to be the date that the Board of Directors lost its quorum as a result of resignations (to be discussed in Chapter 9). Consequently, although the staff could continue to prepare a final assistance agreement, there was no Board to approve it for another seven months at least. Then the new quorum of the Board (which has been referred to as the second Noble Board) wished to consider the terms of the letter (and those of other projects as well) afresh. This process was so convoluted and so intertwined with the events leading to the demise of the Corporation, further discussion of the Great Plains Project will be deferred until Chapter 11.

Solicitation for Coal-Water Fuel Projects

During 1983 the Board was requested by a potential producer of coal-water fuels to issue a solicitation for proposals for such projects (pursuant to Section 127(c) of the Act). This category of technology had not received any considered attention by the Corporation for a number of reasons. First, there was the question as to whether coal-water fuels were truly *synthetic* fuels. Second, the nature of the technology entailed modest scale and cost. Thereby, it was not immediately obvious that the private sector itself could not bear the cost of a commercial facility (which was not an issue with the other technologies being considered by the Corporation). Third, because the Corporation had not received any proposals for such projects under the general solicitations, it had not been evident that the technology was ready for commercial applications.

The request to issue a solicitation forced the Corporation to address the implied questions and determine whether it was appropriate to issue a solicitation. In the end, the staff determined that coal-water fuels were indeed synthetic fuels and that they were ready for commercial application—providing some government assistance was available. The reasoning behind these conclusions required some research, however, and is worth being outlined here.

The question as to whether this fuel was a synthetic fuel arose from the definition in the Act. Section 112(17)(A) defined "synthetic fuel" as any solid, liquid, or gas, or combination thereof, which could be used as a substitute for petroleum or natural gas, and which is produced by a chemical or physical transformation of an eligible feedstock (one of which was coal). In addition, the following Section specifically included mixtures of coal and combustible liquids as a synthetic fuel. It did not, however, mention coal-water mixtures.

The staff, of course, consulted with drafters of the legislation, but did not receive conclusive answers. So, as is often the case, the staff set out to see whether an affirmative answer that could satisfy the SFC lawyers could be found. The Corporation's legal staff analyzed issues such as this to satisfy themselves that if the Corporation were challenged in court or in Congress, a stout rationale would be available.

The logic for an affirmative answer was that the coal-water fuel was clearly a substitute for liquid petroleum and that the grinding of the coal and the chemical additives used to stabilize the mixture could meet the test of the definition as to transformation. In addition, if coal-oil mixtures were specifically included in the definition by the Act, it would be foolish to exclude a fuel that avoided the need for the petroleum portion—ultimately the purpose of everything the Corporation was trying to do.

The other questions were a little trickier. Even though the questions were at this time answered in favor of issuing a solicitation, they were eventually to make the solicitation fail. The staff reviewed technical data from one potential proposer to ascertain that the fuel could be prepared reliably and had reasonable prospects for satisfactory, long-term use in large boilers.

That led to another question: Why was the private sector itself not building the plants, given that they could be constructed for *just* tens of millions of dollars? The answer was essentially that the risks were a little too great and the immediate economic return too uncertain for both potential producers and potential users to proceed absent government assistance. The reasoning was as follows. The fuel could result in erosion of burners and other components of boilers and could require some de-rating of existing boilers. Without full-scale testing of sufficient duration, it would not be possible to reassure potential utility users regarding the equivalence of the fuel and regarding the actual economics.

Such testing was not possible unless a sufficiently large coal-water fuel preparation facility was built to sustain an extended test of the fuel. Potential suppliers were not willing to invest in such facilities unless they had a longer-term contract for fuel to allow them to recover their investment costs. Finally, the utilities were not prepared to enter into such longer-term contracts absent conclusive tests. Hence, there was a "Catch-22" situation that the Corporation could easily resolve with its financial assistance mechanisms.

The last major question for the Corporation to consider was whether the fuel had "programmatic" interest. In other words, was the potential application of the fuel great enough to be of interest to the Corporation. Again, the answer was found to be in the affirmative. Electric utilities in the United States were burning in the neighborhood of 1 million barrels a day of residual oil to produce electricity. A significant fraction of the oil could have been replaced by coal-water fuel. The staff reviewed actual listings of boilers of the major utilities for suitability of conversion, in terms of technical configuration and in terms of remaining useful life, determining that boilers installed in the late sixties and early seventies were the prime candidates. In addition, it was through a similar analysis apparent that a market in large industrial boilers had comparable potential for eventual penetration by the new fuel.

Consequently, on January 5, 1984, the Board approved a Solicitation for Coal-Water Fuel Projects. It contemplated making awards—providing price guarantees only—in two categories: projects producing coal-water fuel for use in industrial fuel burning facilities and projects producing coal-water fuel for use in an electric utility power plant. The minimum scale for each project, determined on a Btu-basis, was 1,000 barrels per day COE for industrial projects and 3,000 barrels per day COE for utility projects. The minimum scales were determined by discussions with potential proposers as to the probable size of commercial facilities. It was apparent that if one were to supply a modern utility oil burning facility, a significantly larger fuel producing facility would be required than would be the case for a facility producing for the typical industrial oil burning facility. Nonetheless, proposers were not restricted in having facilities produce only for the industrial or just for the utility markets. The solicitation was merely establishing competitive categories.

The most unique feature of the solicitation was the requirement linking the fuel producers with the consumers. Under the terms of the Act, the Corporation could provide assistance only for the production of synthetic fuels. In this market, as indicated above, commitments by users were essential in crafting an overall viable enterprise. The solicitation was, consequently designed to provide sufficient assistance to the producer of the coal-water fuel that it, in turn, could offer a subsidy to the eventual user as an incentive to switch fuels. Accordingly, the solicitation required proposers (i.e. producers) to have at least a one-year conditional purchase commitment for at least 50 percent of the proposed project's output before an assistance contract would be signed.

The solicitation process was similar to that of the other modified targeted solicitations. Proposers had first to demonstrate that they were "qualified" before the Corporation would discuss terms of financial assistance.

The standards of qualification were virtually identical to those of the other solicitations. Proposers had to demonstrate their managerial capability, the readiness of the technology for the intended application, a viable financial plan (including evidence that 60 percent or more of the necessary capital was available at the time of the proposal submission), adequate cost estimates, and evidence that the site and facilities would be available. In addition, proposers had to show evidence that the facility could be expected to start operation before July 1, 1987.

Applicants

Eleven proposals were submitted in response to the solicitation.[95] These were not, however, eleven different distinct projects. In a few cases the sponsors were offering options as to the eventual location of a facility by submitting an identical proposal for two different sites. Nonetheless, virtually all of the proposers appeared as though they would be qualified, and the staff felt that the response was more than satisfactory if the goals of the solicitation were to be achieved. The eleven projects are listed below in Table VIII-1.

Solicitation Outcome

Although, as previously discussed, several of the solicitations resulted in no awards to projects for various reasons, the failure of this solicitation was solely the work of the Corporation—or more exactly, the second Noble Board.

The Solicitation's deadline for the submission of proposals was June 15, 1984. Consequently, the proposals were received during the period in which the Board lacked a quorum to transact business (upon Vic Thompson's and Jack Carter's resignations on April 26, 1984, only Ed Noble and Vic Schroeder remained in office). The staff proceeded to carry out the evaluations of the proposals and prepared recommendations for the Board, which would necessarily

APPLICANTS TO THE COAL-WATER SOLICITATION

Project	Sponsor	Technology	Scale
Chesapeake (Virginia)	Dominion Resources, COMCO, ARC-COAL, Bechtel Power	ARC-Coal	1,900 BPD
CLI-Florida	CoaLiquid, Inc.	CoaLiquid	1,500 BPD
FW-Elmwood (Tennessee)	Foster-Wheeler	CARBOGEL	5,173 BPD
FW-Mobile (Alabama)	Foster-Wheeler	CARBOGEL	7,675 BPD
FW-Mobile (Alabama)	Foster-Wheeler	CARBOGEL	15,350 BPD
FW-Savannah	Foster-Wheeler	CARBOGEL	7,675 BPD
FW-Savannah	Foster-Wheeler	CARBOGEL	15,350 BPD
Hillsborough Bay (Fla.)	COMCO, ARC-COAL, & Bechtel Power	ARC-Coal	18,000 BPD
North American CWF (New Jersey)	Babcock & Wilcox, Ashland Oil, & Slurrytech	Co-Al	18,000 BPD
Port Sutton (Florida)	COMCO & ARC-COAL	ARC-Coal	4,125 BPD
Standard BPD (Havens, Ill.)	Standard Havens, Gallagher Asphalt, & Rexnord	Standard Havens	1,000 BPD

Table VIII-1

await the appointment of new members. The staff concluded that at least ten of the eleven proposals should be found qualified. The eleventh proposal's qualification would rest on the Board's interpretation of some issues.

The new Board, even though retaining Noble and Schroeder, was very different in temperament from the prior one, which had issued the Solicitation. (Some of which will be discussed in a later chapter.) With regard to this Solicitation, however, additional doubt arose as to whether it was appropriate for the Corporation to become involved in small projects involving relatively simple technologies—an echo of the central issue that had been extensively reviewed before the Solicitation had been issued. It would be more accurate to say that Tom Corcoran and Paul MacAvoy had such doubts. But since the Board only had five members at this time, any two could block it from action. Noble and Schroeder clearly felt that since the Corporation had issued the Solicitation and called forth effort on the part of proposers, the Corporation was bound to make an award. Reichl appeared to feel that applying the technology to a commercial application would be a worthwhile contribution—albeit small.

Given the division of the Board, the differences were awkwardly resolved. After reviewing the staff's findings, the new quorum decided at the January 15, 1985, meeting of the Board to eliminate all projects that had applied under the industrial applications category on the basis that such small projects could be undertaken by industry whenever the economics looked favorable. Consideration of the proposals under the utility applications category was to continue.

Shortly thereafter, however, the Corporation received a letter from a private firm protesting the Solicitation on the grounds that the Corporation was interfering with the efforts of the private sector in general and its efforts in particular to commercialize the technology. If the Corporation were to make a subsidy to one firm, it could clearly offer more favorable terms to a potential user than could a firm such as the protesting one. The private firm assured the Board in the letter that it would be successful in its efforts absent the Government's interference.

Given the strictures of the Act for the Government not to supplant private initiatives, the Board felt it had no choice but to cancel the Solicitation. Reichl, Noble and Schroeder were doubtful and indicated that the Solicitation should be reconsidered at the end of the year if there had been no substantial progress in the private sector. As it transpired, the Corporation was terminated by year end and no reconsideration took place.

Solicitation for Coal or Lignite Gasification or Liquefaction Retrofit Projects

The third of the modified targeted solicitations was directed to projects that proposed to liquefy or gasify coal or lignite in a "retrofit" situation. The Corporation had just had the benefit of experience negotiating the terms of the assistance commitment with Dow and saw how dramatically the amount of required assistance could be reduced for a given scale of plant if virtually all the supporting infrastructure was in place and did not have to be funded by the Corporation. Consequently, the Board wished to encourage additional applications of this kind. As a practical matter, such facilities were only to be found using coal or lignite. There were no oil shale or tar sands facilities in operation in the United States.

The Solicitation was issued in February 1984 and requested proposals for projects having a scale in the range of 3,000 to 8,000 BOED. Price guarantees were the only forms of assistance offered. Only two proposals were submitted under the Solicitation[96]:

The Sohio Project, which was proposed by Sohio Alternate Energy Company, a wholly-owned subsidiary of The Standard Oil Company (Ohio). The proposal was to add the Texaco coal gasification process to an existing facility in Allen County, Ohio to produce synthesis gas equivalent to 7,134 BOED. The project would have used Eastern bituminous coal as feedstock.

The Syngas Technology Project, which was proposed by Syn-Gas Technology, Inc., a wholly-owned subsidiary of Pennsylvania Engineering Corp. The project would have employed a version of the Kloeckner Kohlegas Molten Iron Bath Process at Sharon Steel's plant in Farrell, Pennsylvania. The process would have used Eastern bituminous coal to produce 3,000 BOED of medium-Btu gas.

These proposals were also received and evaluated during the period that the Corporation lacked a quorum on the Board. The staff's findings to be acted upon by the new Board were that the proposals met the Qualification requirements of the Solicitation.

As was the case for the Coal-Water Solicitation discussed above, proposers under this Solicitation got caught by the changed perspective of the second Noble Board. That Board early on decided it wished to do fewer and smaller projects and only those that expanded the diversity of technology. With that mindset, the Board decided that neither of the projects met their new tests.

The Sohio Project would have used the Texaco gasification process. Since that gasifier was already operating as part of the Cool Water Project, the Board felt that an additional project would simply be duplicative and would not justify the obligation of further Corporation funds. With regard to the Syngas Technology Project, the Board felt that the replicability of the technology would be too limited. In their eyes, there were not enough potential users of a steel-making technology.

Accordingly, the Board voted on February 20, 1985, to eliminate both projects from further consideration for lack of programmatic value.

Solicitation for Eastern Province or Eastern Region of the Interior Province Bituminous Coal Gasification Projects

The next of the modified targeted solicitations was not issued until more than a year later, when the Board once again had a working quorum and that Board had established a clear plan for the projects it wished to fund.

The first order of new business for the restored Board was to try to attract a solid array of technologies to exploit the nation's enormous coal resources. Those resources, as noted, are more than twice the size of the nation's oil shale resources, and those, in turn, represent more than twice the oil in the Middle East. To be sure, the synthetic fuels program of the Act (including the efforts of the Department of Energy) embraced three coal projects already—Cool Water, Dow Syngas, and Great Plains. The first of these two projects employed two different versions of pressurized, entrained-flow gasifiers. Great Plains used perfected versions of the Lurgi fixed-bed gasifier. There were, however, other promising coal gasification technologies that had been developed through the pilot stage and now appeared ready for commercial development. And equally importantly, they appeared to have substantial backing by private sector sponsors.

Three important such technologies were the fluid-bed gasifier being developed by Westinghouse, a fixed-bed, slagging gasifier being developed by British Gas and others, and an entrained-flow gasifier employing dry coal feed being developed by Shell, among others. (The entrained-flow technology mentioned here differed from those already funded in that it promised the use of dry feed rather than coal slurry, thereby improving the overall efficiency of the process by not having to provide the heat needed to vaporize the water in the slurry.)

The solicitation stated that the Corporation intended to make an award in each of the three technology categories mentioned above. As such, each category represented a separate competitive field. All would be governed by the same proposal requirements and solicitation evaluation procedures, but the timing could vary for the dates for the various proposal submissions and for decisions to be made by the Board. The Solicitation's structure differed a bit from the other solicitations, thereby building on the experience gained from them. One such difference was the inclusion of an early step in which potential proposers would provide the Corporation with a Preliminary Qualification Statement, outlining a description of the project , as well as the financial, engineering, and management capabilities of the sponsors. The step was designed to allow the Corporation to discuss the prospects for the projects with the sponsors and to help guide them with the formulation of the more extensive proposal submissions that were to follow.

Thus, the Corporation could gain a practical sense of the readiness of potential proposers to proceed with projects. For example, each of the proposers was asked to indicate what the appropriate timing would be for the submission of the more extensive Project Proposal. Inasmuch as this latter Proposal required the submission of cost estimates consistent with a relatively advanced Level II/III design, the Corporation needed a sense as to when the proposers were likely to be in a position to provide a submission containing the desired information.

No project, however, would be dropped for failing to meet the criteria specified for the Statement. On the other hand, no one could submit a proposal later in the process if they had failed to submit a Statement.

The second step in the solicitation process involved the submission of a Qualification Proposal. The intent of this step was to minimize the amount of effort required of sponsors in proposal preparation, and still allow the Corporation to assess realistically whether the project could meet the requirements set out in the Solicitation and whether the sponsors could assemble the capabilities necessary for executing and operating the project, and whether the project was proceeding on a schedule compatible with the Solicitation. Proposers had to provide financial, engineering, management, schedule and other data. Projects not able to meet the Solicitation criteria to the Corporation's satisfaction would be dropped from consideration after an evaluation of the Qualification Proposals took place.

The third step built into the solicitation process involved the submission of a Project Proposal. The requirements for the Project Proposal were developed so as to allow the Corporation to assess whether a Project would be sound as an industrial undertaking and had the potential for making a cost-effective contribution to the Corporation's program. At the submission of the Proposal, sponsors had to demonstrate that they could contribute equity to the Project equal to 15 percent of the total estimated cost of the Project. In addition, the Project had to be at a Level II/III design.

Two other innovative procedures were incorporated in the Solicitation. To allow more expeditious awards, the Corporation would rank the Project Proposals in terms of their relative programmatic value and their overall relative strength. Thereupon, the Corporation would have the option of negotiating only with the highest ranked project in each competitive category. It could, of course, negotiate with all projects meeting the solicitation's criteria. The other feature allowing accelerated progress was the option for the Corporation, by its sole discretion, to authorize competitive negotiations with all qualified projects after evaluation of the Qualification Proposals. In other words, the Corporation could eliminate the Project Proposal. This feature was obviously intended for circumstances in which there were few proposals and at least one from strong and qualified sponsors.

The deadline for submission of the Preliminary Qualification Statements was July 8, 1985, for all three technology categories. Ten projects submitted these statements:[97]

COGA-1, which proposed to build a 1.1 million tons/year ammonia producing facility at Macoupin County, Illinois using a Fluidized-bed/U-Gas technology. The project was sponsored by a consortium consisting of Coal Gasification Inc., Foster-Wheeler, Norsk-Hydro, Freeman-United Coal Co., Union Carbide Corp., Institute of Gas Technology, and Dean Witter Reynolds, Inc.

Davis Synthine, which proposed to produce 30,000 BOED of synthetic natural gas and liquid fuels at Hornell, New York using entrained flow/ low temperature carbonization and KBW Gasification technology. The project was sponsored by the Davis Synthine Fuels and Coal Corporation.

OHIO I-Coal Conversion, which proposed to build a facility producing both methyl fuel and ammonia totaling about 4,000 BOED at Lawrence County, Ohio, using the Fluidized-bed/ high temperature Winkler technology. The sponsoring group consisted of

Wentworth Brothers, Inc., Energy Adaptors Corp., Hoechst-Uhde GmbH, and David Russack Associates.

Keystone, which had submitted under earlier solicitations with a differing sponsor group and larger scale, here proposed to build a 3,500 BOED facility for the generation of electricity at Somerset County, Pennsylvania, using the Fluidized-bed/ KRW technology. The project was sponsored by the Signal Companies, Inc.

Scrubgrass, which proposed a project for producing 2,900 BOED of gasoline at Venango County, Pennsylvania, using Entrained flow/high pressure GKT technology. The project was sponsored by Scrubgrass Associates (AC-Valley Corporation was the General Partner. See discussion of the project in earlier chapters for a listing of the other members of the sponsoring group.)

Air Products Gasification, which proposed to produce 2,500 BOED of two products, fuel gas and methyl fuel, at Memphis, Tennessee, using a Fixed-bed/ British Gas/ Lurgi Slagging Gasifier technology. The project was sponsored by Air Products and Chemicals, Inc.

Calderon Bowling Green, which proposed to generate 4,100 BOED of electricity and steam at Bowling Green, Ohio, using a Fixed-bed/ Calderon Slagging Gasifier technology. The project was sponsored by Calderon Energy Co.

Virginia Power, which proposed to generate electricity from the equivalent of 6,000 BOED of synthetic gas, at an unspecified site in Virginia, using a Fixed-bed/ British Gas/ Lurgi gasifier. The sponsors included Virginia Power, EPRI, General Electric Co., and Consolidated Coal Company.

H-R International Gasification, which proposed to produce 3,000 BOED of fuel gas at an unspecified site in West Virginia, using the Fixed-bed/ British Gas/ Lurgi Slagging Gasifier technology. The sponsors consisted of H-R International, Inc. and a Slagging Gasification Consortium.

Process Energy, which proposed to produce 1,000 BOED of hydrogen and CO, and 1,000 BOED of fuel gas at an unspecified site in West Virginia, using the Fixed-bed/ British Gas/ Lurgi Slagging Gasifier technology. The project was sponsored by Process Energy Systems, Inc.

As can be seen, five of the projects proposed to use either fluid-bed or entrained-flow technologies, and five proposed to use fixed-bed slagging technology. The first group had to submit Qualification Proposals to the Corporation by September 9, 1985, and the second group by November 15, 1985. Six of the ten submitted Qualification Proposals by the respective deadlines.

Of the first group, the Board found that the Keystone and COGA-1 projects met the qualification criteria of the Solicitation and authorized the staff to proceed to competitive negotiations (one of the options allowed under the solicitation that avoided the requirement for the sponsors to prepare an extensive Project Proposal). Three other projects, Ohio I, Scrubgrass, and Davis Synthine, were found to not meet the requirements of the Solicitation for Qualification.

Of the second group, only Virginia Power submitted a Qualification Proposal by the deadline. The Board did not have an opportunity to act on this proposal before Congress terminated its authority. Based on the staff evaluations, however, the proposal showed considerable promise.

This Solicitation was never completed because of congressional action, but it could well have been one of the most successful of those issued by the Corporation. In the author's estimation, all three of the qualified (or likely to be qualified) projects could have successfully negotiated awards with the Corporation.

Solicitation for Projects to Produce Synthetic Fuels by Mining and Surface Processing Tar Sands

The last of the modified targeted solicitations was the Solicitation for Projects to Produce Synthetic Fuels by Mining and Surface Processing of Tar Sands. This Solicitation was issued on June 24, 1985, in response to a request by a potential sponsor. The request was favorably received by the second Noble Board because its newly formulated "Business Plan" called for at least one tar sands project in order to meet the diversity objectives of the Energy Security Act.

The Solicitation invited proposals only for tar sands projects using technologies for the mining and surface processing of tar sands. There are, of course, significant deposits of domestic tar sands that can be accessed only through in situ recovery techniques. But the Corporation already had a heavy oil project under consideration that involved in situ techniques that could also be applied to a significant number of tar sands deposits. It was a surface mining and solvent extraction technique that was lacking from the array of projects being considered. (As we will see, one such project, Kentucky Tar Sand, received a letter of intent from the Corporation prior to this time, but lost its interest in further development and had withdrawn its application for assistance.)

While the Solicitation still took the general form of a modified targeted solicitation in its focus solely on one technology and resource, its structure differed somewhat from the other targeted solicitations. In effect, the Corporation crafted the Solicitation procedures to match its understanding of the projects likely to be applying under the Solicitation. Specifically, the project, which requested the issuance of the Solicitation, had applied under earlier solicitations. It was clear that some other such projects might apply as well based on staff estimations informed by earlier reviews. The projects involved promising technologies of interest to the Corporation, and included sponsors that had been primarily responsible for developing those technologies.

The projects lacked, however, sponsors with the financial strength that would be required to raise the necessary equity on their own, absent tangible signs that the Corporation was likely to provide financial assistance. In addition, while the projects had done considerable design work for a facility on a site-specific basis, by and large, they were not likely to have designs comparable to the advanced Level II/III generally required by the Corporation at the time of proposal submission. Additional private sector funding would have to be raised to do the additional engineering and design. Again the necessary funding was not likely to be found until the likelihood of Corporation assistance was imminent. That at least was the picture painted by the potential projects, a picture of some plausibility in light of conversations that the Corporation staff had with potential equity investors in the projects.

Accordingly, the Solicitation dispensed with a Preliminary Qualification Statement inasmuch as it was already clear that a number of promising projects were likely to apply. In addition, the Solicitation did not incorporate the two-step proposal—i.e., a Qualification Proposal, followed by a more costly Project Proposal.

Rather, the Solicitation simply called for a "Project Proposal" whose requirements were less stringent on applicants than the Project Proposals of the other targeted solicitations. In particular, the Project Proposal only had to contain a cost estimate comparable to a Level II design. The project, however, would have to have developed a Level II/III design before a financial award was actually made. The concept was that new equity sponsors would then step up to make the necessary investment in design to allow the final deal to be consummated once it was evident that the Corporation liked the project sufficiently to be willing to make an award. Similarly, the Solicitation did not require a substantial demonstration of equity at the time of the proposal. Instead, the Solicitation indicated that it would select projects for negotiations after the evaluation of the Proposals, and then the projects would have a sixty-day deadline to assemble the necessary equity. No negotiations, however, would proceed until the equity was present.

Otherwise, the Solicitation contemplated virtually the same evaluation of projects in terms of strength criteria as the earlier targeted solicitations did in the areas of technology and engineering, environment and socio-economics, sponsor and project management, markets, availability of a project site, and so forth.

The Corporation received four proposals by the Solicitation deadline. Following the staff's evaluation of the proposals, they found that two met the criteria of the Solicitation:

The PR Spring Project, a joint venture of Solv-Ex Corporation and Enercor to be located in the Uintah Basin of Utah. The Project proposed to use a bitumen solvent extraction process developed by Solv-Ex and a bitumen upgrading process of the UOP Aurabon technology. Cogeneration via fluidized bed combustion of upgrader bottoms completed the overall processing concept. The project was sized to produce approximately 4,600 barrels per day of upgraded synthetic crude.

The Sunnyside Project, which was to be located near Sunnyside, Utah, the site of other major tar sands deposits in that state. The project was designed to produce 5,000 barrels per day of syncrude. The bitumen was to be extracted using the Dravo process. The extracted bitumen was to be upgraded to 26 degree API gravity syncrude using Engelhard's ART process. The project sponsor was the GNC Energy Corporation.

The Corporation's staff found both projects to be approximately equally ranked against the criteria of the Solicitation. Accordingly, the Board directed the staff to work with both projects to encourage them to secure the required equity to proceed to negotiations.[98] No further action had occurred, however, by the time Congress terminated the Corporation.

Only one other solicitation was issued by the Corporation, the Fourth General Solicitation. It is more convenient to discuss that solicitation in the context of the events of the last year of the Corporation in Chapter 11. The next Chapter will examine the Program that was emerging at the conclusion of the first Noble Board in the spring of 1984.

IX

Full Steam Ahead

From early 1983 until mid-1984, the Corporation hit its stride. It continued negotiating the letters of intent from the first solicitations, issued the targeted solicitations, and evaluated proposals from the Third General Solicitation. The previous chapter described the outcome of the Targeted Solicitations, while this one describes how the Third Solicitation produced another seven letters of intent, so that by 1984, the Corporation had three assisted projects underway, and had as well a broad slate of projects in negotiation, which represented all the principal synthetic fuel resource bases and technologies.

THE THIRD GENERAL SOLICITATION

The Third General Solicitation, which had been authorized by the Board on August 19, 1982, closed on schedule on January 10, 1983, by which time 46 project proposals had been received (16 of which had not applied under a previous solicitation). Of these, 10 were for coal liquefaction projects, 9 for coal gasification projects, 13 for oil shale projects, 11 for tar sands projects, and 3 for projects to produce fuels by other means. In addition, the North Alabama Project and the Kensyntar Tar Sands Project were transferred from the previous solicitation evaluation process by Board action to establish a set of 48 projects for further evaluation and possible negotiation.[99] The Corporation proceeded to deal with these proposals in accordance with the procedures described in Chapter 6: maturity review, strength evaluation, and negotiation for assistance (producing either a Letter of Intent or an assistance contract).

Evaluation Results

Twenty-four of the projects passed the maturity tests and sixteen of these passed the initial strength review as shown in Table IX-1— Third Solicitation Evaluation Results[100].

THIRD SOLICITATION EVALUATION RESULTS *Part 1*

Project Name/Location	Sponsor(s)	Production	Passed Maturity Test?	Passed Initial Strength Review?
North Alabama (AL)	Kidder Peabody, Santa Fe International, Air Products and Chemical Co., Raymond International, Houston Natural Gas Co., and Peabody Coal Company	Coal liquefaction: 28,139 BPD of methanol	Yes	Yes
Cathedral Bluffs (CO)	Occidental Oil Shale, Inc. and Tenneco Shale Oil Company	Oil Shale; Case I: 11,700 BPD; Case II: 13,500 BPD	Yes	Yes
Parachute Creek, Phase II (CO)	Union Oil Company of California	Oil Shale: 20,000 BPD	Yes	Yes
Dowsyn (LA)	Dow Chemical Company	Coal Gasification: 5,172 BOED medium Btu gas	Yes	Yes
Northern Peat (Maine)	Wheelabrator-Frye Clean Fuel Corporation	Wet carbonization of peat: 2,910 BOED solid fuel	Yes	Yes
Keystone (PA)	Westinghouse Electric Co., Fluor Engineers, Inc. and Amerigas, Inc.	Coal liquefaction: 13,300 BOED medium Btu gas	Yes	Yes
Memphis (TN)	Memphis Light, Gas and Water Division and Foster Wheeler Energy Corporation	Coal Gasification: 8,600 BOED medium Btu gas	Yes	Yes
Chaparrosa Ranch (TX)	Chaparrosa Oil Company	Tar Sands: 5,000 BPD of Bitumen	Yes	Yes
Forest Hill (TX)	Greenwich Oil Corporation	Tar Sands: 1,750 BPD heavy oil	Yes	Yes

Table IX-1

THIRD SOLICITATION EVALUATION RESULTS *Part 2*

Project Name/Location	Sponsor(s)	Production	Passed Maturity Test?	Passed Initial Strength Review?
Sunnyside (UT)	GNC Energy Corporation and Chevron Resources Co.	Tar Sands: 1,500 BPD	Yes	Yes
White River (UT)	Phillips Petroleum Co., Sunoco Energy Development Corp., and SOHIO	Oil Shale: 16,500 BPD	Yes	Yes
World Energy Inc. (WY)	World Energy Inc. and Extractive Fuels, Inc.	Coal Gasification: 1,100 BOED electricity and 175 BOED oil	Yes	Yes
Kentucky Tar Sand (KY)	Texas Gas Development Corporation	Tar Sands: 5,000 BPD	Yes	Yes
Kennsyntar (KY)	Pittston Petroleum, Ward Douglas, and KSA Resources	Tar Sands: 6,349 BPD	Yes	Yes
American Syncrude (KY)	American Syncrude Corp.	Oil Shale: 4,160 BPD	Yes	No
Means Oil Shale (KY)	Southern Pacific Petroleum and Central Pacific Minerals	Oil Shale: 13,440 BPD	Yes	No
KILngas (IL)	Allis-Chalmers Coal Gas Corporation	Coal Gasification: 1,730 BOED low Btu gas	Yes	No
New England Energy Park (MA)	EG&G Inc., Brooklyn Union Gas, Eastern Gas & Fuel Associates, Westinghouse Electric Co., and Bechtel Power Corporation	Coal Gasification: 8,500 BOED	Yes	No
Iron City (OH)	Iron City Fuels, Inc.	Coal Gasification: 2,930 BOED medium Btu gas	Yes	No
Scrubgrass (PA)	The Signal Companies, A-C Valley Corp., C&K Coal Co., Perry Brothers Coal Co., Mountain States Engineers, Parsons-Brinkerhoff, Krupp/GKT, Lyden Oil, Union Carbide, Pyrofax, and Davy McKee	Coal Gasification: 8,600 BPD of gasoline, 1085 BPD butanes, 800 BPD propane	Yes	No
Enpex Syntaro (TX)	Texas Tar Sands Ltd., Superior Oil, Getty Oil and Whittier Interests	Tar Sands: 800 BPD	Yes	No
Cottonwood Wash (UT)	Magic Circle Energy Corp., Deseret Power and Light, and Foster Wheeler Synfuels Corp.	Oil Shale: 8,260 BPD	Yes	No

Table IX-1

LETTERS OF INTENT

Of the sixteen projects that passed the initial strength review, seven were able to negotiate potential assistance packages as documented by letters of intent signed by the Chairman, of these one, Dow Syngas, also negotiated a financial assistance agreement. Letters of intent for Union Oil Parachute Creek Phase II, Cathedral Bluffs, Seep

Ridge, Forest Hill, HOP Kern River, Kentucky Tar Sand, and Northern Peat are discussed here. (HOP Kern's letter stemmed from the Second Solicitation but was revised at this time.) The Dow Agreement is discussed later.

Parachute Creek - Phase II

Union Oil had, as mentioned earlier, already received assistance from the Department of Energy for the construction and operation of its Parachute Creek Oil Shale Project. The administration of the assistance commitment had been transferred to the Corporation when the president declared the SFC operational in February 1982. Union had all along been planning a major commercial oil shale production facility at Parachute Creek of the order of 80,000 barrels per day. In the minds of Union management, the award by DOE to the first retort of 10,000 BOED was merely Phase I. Accordingly, Union applied to the Corporation under the Third General Solicitation for assistance to construct and operate Phase II—an additional 40,000 BOED. (They had also, as noted previously, applied in parallel to the targeted oil shale solicitation, but were found to be unresponsive in light of the solicitation's stipulated ceiling of 10,000 BOED for projects.) In effect, Union was hoping to have the technology up and perfected with the first retort in 1983, before the serious engineering and design for Phase II were launched.

The Phase I project, which had completed construction at this time, was located on Union's oil shale properties near Parachute, Colorado. The feedstock came from the Mahogany Zone of Colorado's Green River formation, the richest and highest-grade deposit in the United States where shale rock has an average yield of 34 gallons of retorted shale oil per ton of shale.

The Phase I project consisted of three major facilities—the mine, the retort and the upgrading facility. The mine would be a vast affair: at full operation it would dwarf your average coal mine (5 to 10 times larger). The structure of shale allows the creation of large caverns (room-and-pillar mines would have heights and widths of 40 to 60 feet). Operations underground involve massive machine diggers, huge trucks, and shale crushers several stories high. No such mine had ever been operated for an extended period, and a significant amount of learning benefit was expected from the project. At design throughput, the Phase I retort alone required the mining of more than 12,000 tons per day of shale.

The retort consisted of the Unishale "B" technology. It is called a "rock pump" because the shale enters the retort from the bottom and a piston pushes the rock up against gravity. Many other technologies have shale enter at the top of the retort and allow gravity to pull the shale down. As the shale rises in the "B" retort it is heated as it encounters hot gases that are blown into the retort from the top. As the shale reaches approximately 900 degrees Fahrenheit, it releases the kerogen, which flows from the bottom of the retort. The retort facility also includes various auxiliary systems, such as the spent shale shaft cooler and seal leg systems for cooling the spent shale, and environmental control systems.

The Mahogany zone shale at Parachute Creek is up in the mountains, more than 1,000 feet above the valley floor. Union decided to locate the retort adjacent to the mine and built a five-acre ledge for the retort just outside the mine. The cooled, spent shale is transported by conveyer to the valley floor where it is carefully contoured and then planted with vegetation.

The upgrading facility is located approximately eight miles down the valley at the main road (where trucks would pick up the refined product). The upgrading is necessary to convert the raw shale oil to a product that can be handled by most refineries. The raw shale oil contains considerably higher levels of nitrogen and oxygen than does crude petroleum; it also includes some shale fines. Further it contains contaminants in the forms of trace quantities of arsenic and iron. The contaminants must be removed before the product would be acceptable to a refinery. The upgrader uses catalytic hydrogenation processes that include fines removal steps. After upgrading, however, the syncrude is a premium feedstock for a refinery that is particularly suitable for gasoline, jet fuel, and diesel fuels.

The Phase I facility was originally budgeted to cost about $460 million, but eventually cost some $200 million more (in 1981 dollars). The entire investment came from Union Oil—i.e., there was no debt associated specifically with the project. If one stops to think about it, this is a vast amount of capital for any single corporation to devote to a project—especially when considering the amounts to be discussed below for Phase II. This dedication to a project and technology could be clearly traced to the long-term personal interest of Union Oil's Chairman, Fred Hartley. He had done research into the technology for recovering oil from shale at an early age, and became convinced that such resources were going to become nationally important when the conventional oil reserves were no longer available to the United States in sufficient quantities.

Phase II of the Parachute Creek Project was intended to produce approximately 42,000 additional barrels per stream day of hydro-treated synthetic crude oil to be built in two equal increments of 21,000 barrels per stream day.[101]

While these increments were to be built at the Parachute Creek site, they were not to be simple expansions of the Phase I facility. The new retorts were, if technically feasible, to use the "Unishale C" technology, which would have advanced the "Unishale B" by extracting more of the energy contained in the raw shale. Specifically, the "C" technology would involve the addition of a fluidized-bed boiler, which would take the hot spent shale from the retort and burn the residual carbon. In addition, boilers would be included in the design to use the heat generated in the retort to save on energy that would otherwise have to be purchased for the operation of the Phase I facility. The letter of intent[102] allowed, however, that Phase II could employ the "B" technology if "C" proved to be not feasible. Unocal intended to build an additional "semi-works" plant to further test the new technology, before bringing the first increment of Phase II on line.

The new retorts were to be located on the top of the mountain where ample flat surface existed rather than adjacent to the Phase I retort on the ledge. While Phase II would entail opening a new mine, it would expand the existing upgrader. The estimated costs for Increments I and II were estimated at $1.8 billion and $1.4 billion, respectively.

Form and Amount of Proposed Assistance

The Corporation signed a letter of intent to provide assistance to the Phase II Project with Union Oil on December 1, 1983, having the following features:

Maximum SFC Obligation: $2.7 billion, the largest proposed assistance level negotiated for any project applying to the Corporation.

Amount of Price Guarantee: Increment I—$1.55 billion with the 'B' retort; alternatively $1.95 billion with the more advanced 'C' retort; Increment II—an additional $400 million.[103]

Form of Price Guarantee: The guarantee price was initially to be $60 per barrel ($67 per barrel if the "C" technology were employed), adjusted for inflation using the Producer Price Index for Finished Consumer Goods Excluding Food after September 1983.

Duration of Price Guarantee: Price guarantee payments were to be available to the project for a period of ten years after Initial Production.

Revenue Sharing with the Government: SFC to share market revenues for 16 years when the market price is above stipulated levels of $67 before price guarantee payout and $31 per barrel after payout (in 1983 dollars).

Also, the letter contained conditions to be achieved by the sponsor before the final contract could be completed. In contrast to some of the other letters discussed earlier, these conditions did not rely on the performance of additional technology work of any consequence. The conditions went largely to the Sponsor completing an Information Memorandum and demonstrating that all of the regulatory permits could be gotten on a schedule needed to meet the stipulated construction milestones.

Cathedral Bluffs

The Cathedral Bluffs Project was the second of the oil shale projects with which the Corporation negotiated a letter of intent.[104] The negotiated assistance represented the second largest amount among all the projects applying to the Corporation and its proposed technology was among the most complex of those evaluated. The partnership sponsoring the project consisted of wholly-owned subsidiaries of Occidental Petroleum Corporation and Tenneco, Inc. Occidental was the developer of the central technology being proposed. Just as Union Oil's steadfast support of its shale development could be traced to the force of Fred Hartley, its chairman, so could Occidental's determination be derived from its forceful chairman, Armand Hammer.

The unique features of the Cathedral Bluffs Project stemmed from its central location within the bowl-shaped Green River Shale formation. Ages ago when the shale hydrocarbons were laid down, the formation was essentially the flat bottom of a sizable body of water. With time, however, mountains emerged around the rim of the formation, forcing the outer perimeter up and the center down—that is in the shape of a bowl. And with the passage of time, an overburden of one to two thousand feet was deposited in the center. While Cathedral Bluffs was located in the center, Parachute Creek, as mentioned, was located at the outer perimeter where oil shale deposits were found close to the top of mountains.

In the prior decade, the Department of the Interior attempted to encourage private corporations to develop means for the recovery of shale oil from the various parts of the formation. To this end they leased substantial tracts of shale deposits to developers on the basis that they were to spend certain sums over time to develop the sites. The sponsors of Cathedral Bluffs were granted the so-called C-b federal lease tract in Rio Blanco County, Colorado. Under the terms of the lease arrangements, the sponsors had already invested in the order of two hundred million dollars in the site before the project applied to the Corporation for assistance. This was obviously not a game that every corporation could play.

Occidental's challenge was to find a technology that had the promise of processing the deep shale deposits economically. It became clear that an attempt to mine shale fifteen hundred feet under the surface and bring it to an above-ground retort would never be economically competitive with other projects able to mine equally rich shale at, or close to, the surface. This led to the development of the *modified in-situ* shale technology.

Modified in-situ (MIS) entails development of in-ground retorts in which fire fronts drive out the kerogen from the shale, which is collected and piped to the surface, supplemented by above-ground retorts. Before a fire front can take place, the solid ground must first be broken up to allow the passage of flames. To that end, an underground retort's boundaries are established, the top twenty percent of the shale is mined by conventional means, and then, through the use of explosives, the rest of the shale in the retort is rubblized. Thereby, the mined volume is filled up by the expansion of the exploded shale below. Since the mined shale must be brought to the surface, it makes economic sense to included "conventional" surface retorts to process this shale into crude shale oil as well.

The major technical development was determining how to control a fire front driven by air injection vertically upward through the retort in a way that would optimally recover the kerogen contained in the shale (remember that this is going on fifteen hundred feet underground). When the Corporation's technical staff first reviewed the project, they were not unanimous that the technology was ready for commercial development. During this period, however, Occidental was conducting extensive tests at their Logan Wash site where such retorts could be built and tested closer to the surface at lesser development costs. The results of these tests showed that substantial recovery of the shale oil could be obtained with reliable and predictable results.

The Cathedral Bluffs Project was designed to produce approximately 14,100 barrels per calendar day of a commercial crude oil substitute using the modified in situ technology. In this application, the technology would have consisted of one above-ground retort with a design capacity of about 11,000 BOED, and four in situ retorts (at any one time) producing about 2,300 BOED. An in situ retort would operate the better part of a year in extracting the raw shale oil. So, the in-situ operation would require an ongoing sequencing of the retorts, with new caverns being prepared while the ready caverns were being fired.

Because the amount of shale mined to prepare the in-situ caverns would be insufficient to allow the above ground retort to operate continuously, a commercial room-and-pillar mine would also be operated to provide shale to the retort. The retort technology was to be Union Oil's Unishale B Process. The crude shale oil from all the retorts would be upgraded using Union Oil's upgrading technology on the C-b tract. In addition, the project would have a 46-mile pipeline and terminal facilities that would allow the up-grade fuel to be shipped to Rangely, Colorado. The expected production life of the project was expected to be not less than 30 years, during which time the project was expected to produce about 150 million barrels of product.

Because of the terms of the Department of Interior's leasing program, construction of the project had already begun. Indeed, the sponsors had spent on the order of $250 million on facilities largely associated with the mine, such as shafts down to the fifteen hundred foot level as well as the associated elevator and ventilation systems. Mechanical completion of the above ground retort and the upgrading facilities were expected about four years after the signing of a final Commitment. The in situ retorts would begin operation approximately six months later. The estimated costs of the entire facility (including the sunk costs) were of the order of $2.665 billion.

Form and Amount of Proposed Assistance

The letter of intent included the following features:
Maximum SFC Obligation: $2.19 billion
Amount of Loan Guarantee: $1.812 billion
Amount of Price Guarantee: An initial amount of $378 million, with additional amounts available as the loan was repaid.
Form of Price Guarantee: The guaranteed price for the syncrude was $60 per barrel, adjusted for inflation after March 1983.
Duration of Price Guarantee: The Corporation's obligation to make price guarantee payments would terminate ten years after the Start of Production.
Revenue Sharing with the Government: The project was to share revenues with the Government for a period of 16 years after the "Cut-Off Date" (the date after which no more guaranteed debt could be drawn down).

The letter of intent contained a number of conditions to be met by the sponsors before a final Commitment was signed. While most were procedural in nature, two were substantive. First, an additional equity sponsor had to be brought into the project. In contrast to other projects, this provision did not reflect the lack of committed equity, but rather was designed to avoid voting impasses among equity holders that could easily occur with Occidental and Tenneco, each owning exactly 50 percent of the project. The second condition of note was that the costs of the project would have been audited to the satisfaction of the Corporation for eligibility (particularly the amount of sunk costs claimed by the sponsors in the capitalization of the project and in the calculations of required additional equity).

Seep Ridge

The Seep Ridge Shale Oil Project represented a different technology than the other two shale projects described above. As discussed previously, some shale resources were significantly underground (as at the C-b site) or high in the mountains (as at Parachute Creek). Other resources, of course, lay near the surface. Some projects that did not apply to the Corporation intend someday to mine that shale directly with open pit mining techniques.

Seep Ridge, on the other hand, proposed to extract the shale oil using true in situ techniques. The reader may recall that Cathedral Bluffs proposed MIS techniques because the great depth of the resource necessitated the mining excavation of some of the shale caverns before the shale could be rubblized. In the case of resources that lie near the surface, such mining preparation in advance of rubblization is not required. One can simply blast the shale with explosives. The ground rises a bit with little consequence. Then, the shale is ignited in the ground and a horizontal flame front moves through the ground to create hot gases that cause the kerogen to be released. The raw shale oil is then pumped to the surface.

The project was sponsored by the Seep Ridge Shale Oil Company, a partnership composed of Geokinetics, the developer of the technology, and Gilbert Shale Oil Company, a single purpose subsidiary owned by Peter Kiewit Sons, Inc. Kiewit would be the construction manager and operator.

The Seep Ridge Project was to be located southeast of Vernal, Utah. The shale resources in Uintah County represented a different

shale resource from those to be used from the Green River formation by the Parachute Creek and the Cathedral Bluffs projects. The
Utah shales were also extensive in amount, albeit somewhat less
rich than the Colorado shales (i.e. about 24 gallons per ton as opposed to 34 or so gallons per ton). The project was to produce approximately 1,000 barrels per day of raw shale oil. The raw shale
oil would be sold to refiners in the Salt Lake City area to blend and
sell as fuel oil. The in situ technology was named by Geokinetics
LOFRECO, standing for "low front end cost." A major virtue of the
in situ technology was that it did not require the substantial front
end capital investment in mines and retort facilities that were required by the other shale projects discussed above. Rather, the true
in situ technology resembled conventional oil field development in
its economics. The project's development would be characterized
by the sequential drilling of wells for explosives and for raw shale
extraction. One in situ retort, producing approximately 100-200
barrels per calendar day with a limited lifetime, would be followed
by another.

Specifically, Seep Ridge expected to have seven two-acre retorts
operating at a time to achieve the design production of 1,000 BOED.
Among the project's support facilities would be a 10-megawatt
power plant to provide electricity to the facility using the off gases
of the underground combustion for fuel. The sponsors controlled
sufficient shale reserves to allow the project to operate for at least
20 years (in two adjacent sites).

The total costs estimated for the construction and startup were
$37.3 million, which included $1.5 million incurred as prior costs
by Geokinetics.

Form and Amount of Proposed Assistance

The letter of intent[105] for the project included the following
features:
Maximum SFC Obligation: $45 million in loan and price
guarantees.[106]
Amount of Loan Guarantee: $21.3 million.
Amount of Price Guarantee: $23.7 million, with additional
amounts available as the loan was repaid.
Form of Price Guarantee: $42.50 per barrel of crude shale oil

in third quarter 1983 dollars, escalated thereafter using the PPI.

Duration of Price Guarantee: 10 years.

Revenue Sharing with the Government: Cumulative after-tax cash flow sharing with the SFC when greater than $10 million for 15 years after the start of production.

The letter of intent also contained a number of conditions to be met by the project sponsors before the Corporation would consummate a final commitment. Most were administrative in nature—i.e., demonstrating matters already largely existing. Two of the more substantive conditions required that: (1) the SFC shall have audited and found satisfactory the development data and operating costs, as well as the product quality and yield from the project's last two operating retorts, and (2) that the sponsors have adequately justified the sunk costs that were to be recognized as part of the sponsor's equity in the project.

Forest Hill

The Forest Hill Project was one of the few heavy oil projects that reached an advanced stage of consideration by the Corporation. As discussed above, in connection with the Calsyn Project, not all heavy oil resources were eligible for assistance from the Corporation. In order not to assist heavy oil projects that could be viable on their own, or at least compete with similar projects that were solely funded by the private sector, the Act included a test that the heavy oil required the use of a new technology, whose cost and risks were comparable with the other synthetic fuels technologies, and that its recovery would not be commercially viable absent the assistance from the Corporation. This test required that the staff carry out a technical and economic comparative analysis to establish eligibility. It did so, and the project was determined to be eligible.

The sponsor of the project was Greenwich Oil, the principal stockholders of which are: Union Carbide, Core Laboratories, Inc., Lubrizol Enterprises, Inc., T.N. and C.W. Law, Euclid Partners, Company officers, and others. Several of these had funded a small pilot project proving the concept of the proposed technology on the same site proposed for the full-scale project.

The Forest Hill Project entailed the production of heavy oil from the Harris Sand Reservoir at a site in Wood County, Texas. The site is near the small town of Quitman, about two hours northeast of Dallas. The technology to be used was one that employed oxygen fire-flooding as the means for driving the heavy oil from the Harris Sands, making it possible to bring the oil to the surface by pumping diluent into the ground and raising the mixture to the surface with conventional pumping. The features of the technology that were new—and of programmatic interest to the Corporation—entailed the use of 90 percent pure oxygen being pumped into the ground, as well as the one mile depth at which the fire flooding takes place. This had not been done elsewhere and entailed significant technical risk. To be sure, other projects had used air-driven firefloods for the secondary recovery of oil, but these were not subject to the same degree of risk. The use of the oxygen was required if the techniques were to become commercially viable because of the substantial depth of the resource. Economically, it was not feasible to pump air—with its 80 percent nitrogen content—for this purpose because roughly five times the volume of air would have to be pumped to achieve the same degree of combustion.

The purpose of the combustion was several-fold. It, of course, generated considerable heat, which reduced the viscosity of the heavy oil. In addition, the combustion gases particularly, the CO_2 —would be absorbed by the oil, again reducing the viscosity. At the same time, the combustion would increase the pressure, thereby driving the heavy oil from the injection wells to the producing wells.

In examining the replication potential of the technology, the SFC staff estimated that on the order of 2 billion barrels of heavy oil existed in similar domestic reservoirs for which existing technology would not be economically suitable. A comparable amount of tar sands for which the technology would be applicable appears to exist as well. While these amounts are not large in comparison with the shale and oil resources of the nation, they would have been a respectable addition to the 26 billion barrels of conventional oil reserves that were estimated at this time.

The assisted commercial project was to build on existing pilot work at the project's Wood County site. The sponsors had leased the 2,000-acre site and the right to produce from the Harris Sand Reservoir underlying that site. The reserves were estimated to be sufficient to allow the project to operate at design capacity in excess of 24 years.

The commercial project was to include 25 oxygen injection wells (including the refurbishment of several existing wells from the pilot operation) and approximately 100 production wells. In addition, the project would have attendant surface facilities for gathering the produced oil and for separating water, gas and sediment from that oil in preparation for shipment to the purchaser, as well as for injecting the diluent necessary for bring the heavy oil to the surface.

An essential element to the project was the oxygen production facility. It was to be constructed and wholly owned by the Linde Division of Union Carbide, the major partner, but its costs would not be included as part of the cost of the project in the financing of the guaranteed debt. The oxygen plant's air separation facilities would provide up to 150 tons per day of oxygen. Until the oxygen plant would be completed, Union Carbide would truck in liquid oxygen to support the early fire flood, which would have to precede production by the better part of a year. Apart from the oxygen facility, the project would not appear at the surface to be significantly different from other nearby conventional oil fields.

When the project reached design capacity, it would produce approximately 1,750 barrels per day of 10-degree API oil. This level of production, however, would be achieved over a period of a couple of years as the wells were drilled and as the effects of the oxygen injection and fireflood were translated into higher temperatures and pressures underground. The total cost of the facility was estimated to be $42.077 million, of which $10.077 million represented prior costs expended by the sponsors at the project site.

Form and Amount of Proposed Assistance

The assistance package outlined in the letter of intent[107] contained a combination of loan and price guarantees.
Maximum SFC Obligation: $60 million.
Amount of Loan Guarantee: $22.57 million (including $1.87 million allowance for accrued interest). This amount

corresponded to about 65 percent of the cost required to complete the commercial facility. The other 35 percent was to be derived from contributed equity by the sponsors ($6 million) and from internal cash generation ($5.3 million).

Amount of Price Guarantee: $37.43 million to increase as the loan was repaid.

Form of Price Guarantee: The Guarantee Price was to be $40 per barrel for the first 500,000 barrels of Eligible Product and $37.50 per barrel thereafter.

Duration of Price Guarantee: The earlier of 10 years or the maximum aggregate payment of price guarantees.

Revenue Sharing with the Government: For a period 16 years after a "Cutoff Date", the project payments to first prepay the guaranteed loan, and thereafter make profit sharing payments to the SFC.

The letter of intent contained the usual conditions that had to be met by the sponsors before a final agreement could be consummated. Several additional substantive conditions were included that were unique to this project. The sponsor had to deliver to the Corporation satisfactory evidence of the net value of the existing assets of the borrower and the total estimated costs of the project had to be verified by an independent cost assessment. In addition, Carbide's initial investment had to have been made. Finally, the Term Loan (a loan made by the sponsors to the project prior to the application to the SFC) had to be restructured on terms satisfactory to the Corporation.

HOP Kern River

The HOP Kern River Project was a second heavy oil project entailing the use of in situ recovery techniques to receive a letter of intent from the Corporation. Indeed, chronologically, this project should have been discussed earlier inasmuch as a term sheet had already been negotiated under the competition of the Second General Solicitation. Events, however, were such that the term sheet was not extended to a letter of intent approved by the Board at that time. It was not until early 1984 that the Board authorized a letter of intent to be signed, and, then under renegotiated terms.

The in situ technology proposed for the HOP Kern River Project was entirely different from the technology proposed by the Forest Hill Project discussed previously. The HOP Kern technology was a cycle and steam-drive recovery technique particularly suited for the heavy oil deposits in that part of California. These deposits consisted of highly viscous oil that could not be recovered economically by use of conventional oil drilling. In addition, they lay so close to the surface that secondary steam flooding techniques that had worked in other locations to reduce the viscosity of oil were not practical because much of the steam would escape to the surface.

Consequently, one of the project sponsors, Heavy Oil Process, Inc., a wholly-owned subsidiary of Ladd Petroleum, developed a technological approach that involved the excavation of a shaft into the oil formation to be followed by the drilling of horizontal tunnels into the formation that would allow the efficient injection of steam into the reservoir.

This new approach had been tested in one pilot unit on the western boundary of the proposed project site, which consisted of 280 acres being leased from Shell Oil Company at the periphery of the Kern River field. The pilot unit consisted of a subterranean cavern, lateral wells radiating there from, vertical steam injection wells, and the associated surface equipment necessary for operating the unit (steam generator, pumps, buildings, and monitoring equipment). The heavy oil deposit to be developed contained approximately 22 million barrels of oil at a depth of 250 to 500 feet. The project was to consist of another three units, each consisting of two production patterns, and a cogeneration element, as well. This expansion project was expected to produce on average over a ten-year period 3,891 barrels per day of oil. The three new units would be constructed and would come on line at one and a half year intervals, with production from the first commencing in November 1987. Each unit would have an operating life of approximately eight years.

Form and Amount of Proposed Assistance

The letter of intent[108] for the project contained a combination of loan and price guarantees:

Maximum SFC Obligation: $100 million.

Amount of Loan Guarantee: $71.2 million (including interest allowance of $6.0 million).[109]

Amount of Price Guarantee: $28.8 million to increase as the loan was repaid up to a maximum of $100 million[110].

Form of Price Guarantee: $28.50 per barrel adjusted monthly for inflation after 1984 by the Producer Price Index for Finished Consumer Goods Excluding Food.

Duration of Price Guarantee: 10 years.

Revenue Sharing with the Government: After cumulative after tax cash flow reached $25 million, the sponsor was required to make mandatory prepayments of the guaranteed debt or if the debt were already extinguished, share profit with the SFC.

The HOP Kern River Project letter of intent also contained a number of conditions that had to be satisfied before the Corporation would be willing to complete a formal assistance agreement. Among the more substantive were the requirements that the project's estimated costs be verified by an independent cost estimate and that the Borrower continue to operate Unit I and thereby confirm the technical feasibility of the project. The latter requirement turned out to be particularly significant. To be sure, Unit I had already produced heavy oil using the new technology. The key question, however, was whether the technology would result in the percentage recovery of oil in place that would be required for economic operation of the facility. The Unit had not been operated at capacity for a sufficient duration up to this point.

Kentucky Tar Sands

The Kentucky Tar Sands Project was the second project proposing to mine tar sands and separate the oil using a solvent extraction process that received a letter of intent from the Corporation (the

Santa Rosa Project being the first). This was the only such project in the eastern United States that achieved an advanced stage of consideration by the SFC.

The proposed project consisted of contracted surface mining of tar sand deposits in Logan County, Kentucky, and a tar sand extraction plant using the Dravo solvent extraction technology. The project was to produce approximately 5,000 barrels per day of 11-degree API crude oil. The sole sponsor of the project was the Texas Gas Resources Corporation, which was a wholly-owned subsidiary of the CSX Corporation. In turn, a subsidiary of the former, the Elute Corporation, was to own and operate the plant.

The tar sands deposits underlay an approximate 2,250 acre site leased from various parties by a joint venture owned 50 percent each by Texas Gas and Green Construction of Indiana, Inc. The latter was to mine the tar sand deposits and transport ore to the plant and take waste material to the mine reclamation areas under a long-term contract with Elute. The dedicated reserves would have been sufficient to operate the project in excess of 25 years at the designed capacity. In addition, the joint venture had under lease additional reserves that would have allowed the expansion of the project to 15,000 barrels per day. The sponsor had previously constructed and operated a pilot plant incorporating the Dravo solvent extraction process to determine engineering performance estimates for the proposed project.

The financial assistance envisioned in the letter of intent[111] consisted solely of a price guarantee, since Texas Gas would have financed the entire construction and operation of the facility on its own. Such arrangements were feasible where the estimated cost of construction was relatively modest and where the sponsors had a great deal of confidence that the technology would work at design levels without an extensive startup period. In this case, the estimated cost was $150 million, a sizable sum, to be sure, but still only a fraction of the larger projects being considered by the Corporation.

Form and Amount of Proposed Assistance

The terms of the letter were:
Maximum SFC Obligation: $543 million in price guarantees.
Form of Price Guarantee: $55 per barrel for the first 3.3 mil-

lion barrels and $40 per barrel thereafter. In each case, the figure would be adjusted quarterly for inflation after March 1984 using the Producer Price Index for Finished Goods Excluding Food.[112]

Duration of Price Guarantee: The earlier of ten years from the date of initial production or maximum aggregate payment of price guarantees by the SFC.

Revenue Sharing with the Government: For a period of 25 years after initial production, the project would make payments to the government when its cumulative after tax cash flow exceeded $150 million, but the payments could not exceed the amount of price guarantees received nor $271.5 million.

The maximum obligational authority was relatively large compared to the cost of the facility. The reader should keep in mind, however, that the "project" receiving assistance consisted only of the solvent extraction unit, while much of the costs of the mining and transportation were not being borne by the sponsor. Nonetheless, the income stream to the project—being supported by the price guarantees—had to be sufficient to pay the costs of the long term contract with the mining operation and, through those payments, eventually pay off the associated capital costs.

The letter of intent also contained some conditions to be met by the sponsors before the final commitment was to be consummated, but these were the procedural kind common virtually to all the letters. There were none linked to specific uncertainties or major development tasks to be performed by the sponsor.

Northern Peat Energy

The letter of intent negotiated with the Northern Peat Energy Project was the second such agreement that the Corporation drafted with a potential peat project, the First Colony Project being the first.[113] The Northern Peat Project was, however, the first project to receive a letter of intent to produce a solid fuel, namely solid pellets of peat-derived fuel (PDF). PDF was designed to be a substitute for residual and distillate fuel oils in industrial, utility, and commercial boilers throughout New England.

While solid peat fuels were clearly not in the same category as shale and coal in their potential as alternative fuels, peat was undoubtedly an eligible feedstock pursuant to the Act, and solid fuels were eligible products under the Act as well. The question for the Board was whether the size of the national peat resource was sufficient to fund the project with an objective of achieving the Act's goal of diversity of feedstock and technology. The staff had researched this question in terms of the standard replication analysis done for every project under serious consideration. It concluded that the peat reserves were sufficient to justify a project, and, given the paucity of other energy resources in the New England area, it was worthwhile to support this project.

The Northern Peat Energy Project was to be located on 300 acres in Milford and Bradley, Maine. The PDF would have been produced using the Peat Wet Carbonization Process, a proprietary process developed by J.P. Energy Oy of Finland. The project would have produced 290,000 short tons per year of clean, low sulfur, low ash PDF, equivalent to 2,745 barrels of oil per calendar day. The process required that peat be harvested wet by floating dredges and pumping it to a dewatering plant where the peat would be heated, filtered, dried and compacted into fuel pellets. The peat feedstock would be harvested from the Sunkhaze Bog in Milford and from the Chemo Bog in neighboring Bradley. These bogs have sufficient peat to operate the facility in excess of twenty years at the designed capacity.

The project was to include: (1) peat extraction and transportation equipment, (2) feedstock-slurry storage basin, (3) process equipment and buildings, (4) rail and truck load-out facilities, (5) administration and laboratory building, (6) wastewater treatment plants, and (7) steam plant and related plant utilities, and certain other structures and ancillary equipment. The direct and indirect construction costs (including contingency and escalation) were estimated at $175.2 million. When interest expense and working capital needs were included, the total eligible costs were expected to be $228.4 million.

The project would be owned and operated by the Northern Peat Company, a wholly-owned subsidiary of Signal Energy Systems, Inc., itself a wholly-owned subsidiary of the Signal Companies, Inc.

Signal, during this period, was an active sponsor of several other major projects being considered by the Corporation for assistance — i.e., Paraho Ute and Keystone.

Form and Amount of Proposed Assistance

The letter of intent contained a combination of loan and price guarantees:

Maximum SFC Obligation: $365 million.

Amount of Loan Guarantee: $160 million including an allowance for accrued interest of $ 13.8 million.

Amount of Price Guarantee: Initial amount of $205 million to be increased to a maximum of $365 million as the loan is repaid.

Form of Price Guarantee: $10.50 per million Btu of product as adjusted monthly for inflation after February 1984 by the non-seasonally adjusted Producer Price Index, excluding food.[114]

Duration of the Price Guarantee: Ten years from initial production.

Revenue Sharing with the Government: After cumulative after tax cash flow reached $35 million, the sponsor was required to make mandatory prepayments of the guaranteed debt or if the debt were already extinguished, share profit with the SFC.

The amount of the loan guarantee was determined by the estimated cost of the project and of the financing plan proposed by the project sponsor. Because of a significant amount of work that would be required to satisfy permitting agencies, the funding structure had a unique upfront feature. Namely, for the first $11 million of expenditures, the eligible costs would be funded equally by guaranteed debt and new equity contributions. Then, after the SFC was satisfied that all the permits could be obtained, the remainder of the guaranteed debt could be drawn down in the proportions corresponding to the Financing Plan. In effect, the SFC would be sharing in the cost of pre-construction work such as could have been supported by grants to refine cost estimates authorized by

Section 131 (u) of the Energy Security Act. In this instance, the SFC decided to wrap the costs in the assistance commitment rather than making it a two-step process. More importantly from the sponsors' perspective, they knew that if they invested the $5.5 million and got all the permits, they had a done deal and did not have to confront any uncertainty of whether the Corporation would ratify the entire guarantee commitment.

The letter of intent contained a number of conditions that had to be met prior to closing the Commitment that were similar in nature to the other letters. The more noteworthy of these were fourfold. The sponsor had to provide evidence that it had paid not less than $2.8 million for contractually eligible property and services that it wished to be credited to its share of the project equity. The sponsor was to have provided evidence satisfactory to the Corporation that could obtain all permits and regulatory approvals in a realistic time frame consistent with the construction and operation schedule for the project. While this last condition was contained in most of the letters of intent, as discussed above, in this instance it had real teeth and would require substantial work to demonstrate. Next, the borrower had to complete a characterization of the plant site that was adequate to substantiate the Level II/III design and cost estimate. This would also entail non-negligible expenditures. In addition, the borrower was to have executed a marketing agreement that contained the substance of the terms contained in a letter agreement between the borrower and C.H. Sprague and Son, or have entered into a comparable alternative marketing arrangement. After all, the PDF was not a usual commercial commodity.

FINANCIAL ASSISTANCE AWARDED TO DOW SYNGAS

Negotiations proceeded so rapidly with the sponsors of the Dow Syngas that a letter of intent[115] signed on February16, 1984, led in short order to a final assistance agreement[116] by April 26, 1984.

The Dow Syngas Project encompassed the design, construction, and operation of a facility for converting coal into medium-Btu synthetic gas. The facility, owned by Louisiana Gasification Technology, Inc. (LGTI), a wholly owned subsidiary of the Dow Chemical Company, the project sponsor, is located near Plaquemine, Louisiana in the Iberville and West Baton Rouge parishes, within an existing Dow petrochemical complex. The project's principal technology is a Dow-developed coal gasification technology particularly designed to use low ranked coals, such as western sub bituminous and lignite coals, in part because Dow held substantial undeveloped lignite reserves that could eventually be used by the gasifier. Inasmuch as substantial costs would have been required to develop mines on the lignite resources, however, the project during its initial years of operation was to use sub bituminous coal transported by rail from the Powder River Basin in Wyoming. The technology is also capable of using higher ranked bituminous coals.

At full capacity, the facility was designed to operate with a sub bituminous coal feed rate of about 2,400 tons per day (tpd), or 3,000 tpd of lignite, to produce 30 billion Btu per day of fuel gas (consisting principally of hydrogen, carbon monoxide, and carbon dioxide) having a heating value of 200 to 260 Btu per standard cubic foot. At full capacity, the project was also to produce about 2,467 tons of 475-psia steam, 17 tons of sulfur, and 255 tons of slag per day. The fuel gas was to be used in existing combined-cycle gas turbines for co-generating electricity and steam; the steam to be exchanged with other Dow plants within the Plaquemine complex for lower pressure steam and power to be used by the facility. Sulfur was to be sold in the local market; slag was to be sold locally or be used by Dow's Plaquemine operations for structural landfill.

The facility consisted of the following process units: coal grinding and slurry preparation, gasification and high-temperature heat recovery, gas cleanup to remove particulate matter, GAS/SPEC ST-1process for hydrogen sulfide removal, and Selectox process for sulfur recovery followed by the burning of tail gases. These units are integrated into previously existing power generating units and other infrastructure (e.g. wastewater treatment, process water treatment, etc.).

In May 1985, a year after the signing of the PGC, Dow submitted a request to the Corporation to include a spare gasifier in the project. This request was predicated on a study that found that a second gasifier was necessary to ensure that the project would achieve its 85 percent design availability. The spare gasifier would enhance on-line availability but would not increase production capacity.

The cost of the second gasifier could be funded from the contingency amount contained within the definitive cost estimate. Moreover, inclusion of the spare gasifier would not delay the schedule for the overall facility and would have no effect upon the Corporation's price guarantee payment obligations to Dow. Accordingly, the Corporation approved Dow's request to include the second gasifier on July 30, 1985.

The oxygen plant was to be constructed and operated by Air Products, a third party supplier. LGTI would purchase the oxygen from Air Products, much as it would purchase other services required by the facility (including power, water, nitrogen, and processing waste water) from the Louisiana Division of Dow.

The definitive cost estimate developed in May 1985 for the design, construction, and commissioning of the plant was $81 million. Considering development work and commitment of working capital, Dow's direct investment in the plant would be approximately $105 million. As noted above, the project uses a number of existing facilities at the Dow complex and purchases oxygen across the fence. Had the project been a stand-alone grass roots facility, the capital investment would have been in excess of $300 million. Thus, the project was a clear example of the type of "retrofit" project the Corporation was seeking to minimize total incremental investment (and eventual government subsidy) in spurring the development of a given technology.

The Dow gasifier employed pressurized, slagging, entrained-flow features, which offered the same improvements as the Texaco gasification technology over other available technologies (see the Cool Water discussion). The technology, however, embodied some distinct differences that were aimed processing low ranked coals reliably and economically. The Corporation's programmatic analysis indicated significant replication potential for sub bituminous coal, which is a domestic resource with more than 400 billion barrels of oil equivalent. The technology was also capable of processing lignite and bituminous coals that are available in comparably large quantities.

Because the project shared pollution control equipment already in place at the Plaquemine site, fewer permits were required than would have been the case for a stand-alone facility. Dow obtained

the necessary additional permits from state and federal agencies
and gained the necessary approvals for renewal of existing permits.

The project's Environmental Monitoring Plan was prepared by
Dow in accordance with the Corporation's Guidelines and was reviewed by the DOE, the EPA, and the Office of Environmental Affairs in the Louisiana Department of Natural Resources. The EMP
was accepted by the Corporation on March 29, 1985.[117]

The Corporation and Dow Chemical letter of intent to complete
a Price Guarantee Commitment (PGC), which would provide financial assistance by the Government to Dow for the construction
and operation of the gasification facility, was described above. The
letter was followed in unusually short order by a completed PGC,
which was signed by all parties on the 26th of April that year.

The Price Guarantee Commitment

The terms of the PGC had the following features:
Maximum SFC Obligation: $620 million in price guarantees.
Form of Price Guarantee: Initially $12.50 per million Btu
    of eligible product for the first 10,950 million Btu and
    $11.00 per million Btu thereafter. The dollar amounts
    are adjusted quarterly for inflation after June 1983. The
    index of adjustment for inflation consists of three fac-
    tors: (i) 30 percent of the increase in the Rail Cost Ad-
    justment Factor, (ii) 35 percent of the increase in the
    Producer Price Index for Capital Equipment, and (iii) 35
    percent of the increase in the Fixed Weighted Price In-
    dex for Gross National Product[118].
Duration of Price Guarantee: The earlier of 10 years or the
    maximum aggregate payment of price guarantees as
    follows:
Primary price guarantee payments not to exceed in the ag-
    gregate $490 million.
Secondary price guarantee payments not to exceed in the
    aggregate $130 million: these not to exceed $40 million,
    $35 million, $30 million, and $25 million in the four suc-
    ceeding 12-month periods.[119]
Revenue Sharing with the Government: For 16 years after

the date of initial operation the project was to make revenue-sharing payments monthly if the market price exceeded the guarantee price or if, after a specified transition date, revenues exceeded a defined level. Aggregate Revenue Sharing Payments were limited to the greater of (i) $310 million or (ii) the aggregate amount of Price Guarantee Payments received by Dowsub (adjusted for inflation using the PPI).

In arriving at the above assistance package, the projected rate of return, market price, and probability of abandonment were major considerations in estimating the minimum amount necessary to induce Dow to proceed with the project. The Corporation calculated that the project would have in the median case (considering the variables for which projections had to be made) about an 18 percent real rate of return on Dow's equity investment of $105 million.

One unusual feature of the Dow PGC vis-à-vis the other price guarantee arrangements was that the payments for each month were to be based on Dow's best estimate in advance of how much product would be produced. All the other project arrangements made payment on the basis of the product actually produced in the prior month. While this was somewhat administratively cumbersome, it speeded up the cash flow to the project and the amount of obligational authority that had to be committed by the Government was reduced to reflect the time value of the money.

Snapshot April 1984:
Considerable program momentum

Despite the initial delays in launching its program, by April 1984, the Corporation could demonstrate substantial progress and credible momentum. There were four projects under construction—i.e., Parachute Creek, Cool Water, Dow Syngas, and Great Plains[120], and eight Letters of Intent either authorized or signed. In addition, the Board had issued its Basic Business Plan (to be discussed in the next chapter), which provided a road map to project sponsors as to its future resource and technology priorities in completing the

Phase I Program, and had in parallel issued the Fourth General Solicitation, which was intended to complete the Corporation's Phase I efforts under the Energy Security Act. Table IX-2—Project Status, April 1984, provides a project summary at this point in time demonstrating the promising momentum developed by the Corporation over the prior year and a half.

Reaching agreement on the above letters of intent along with the potential agreement with Great Plains under the Targeted Solicitation, and the two final Commitments with Cool Water and Dow, all in the space of about nine months, represented convincing progress by the Corporation. This progress showed that the SFC's work with project sponsors had been productive leading them to make the substantial commitments of equity needed for a successful program. Equally important, these projects, along with those

PROJECT STATUS, APRIL 1984

Project	State	Resource/Technology	Status*	Production Rate BPD	Ultimate Site Potential
Cool Water	CA	Coal Gasification Combined Cycle	FA	4,300	8,600
Dow Syngas	LA	Coal Gasification Combined Cycle	FA	5,172	20,000
Great Plains	ND	Coal Gasification SNG Production	LI	23,000	92,000
HOP Kern River	CA	Heavy Oil/ In-Situ Production	LI	3,938	3,938
Forest Hill	TX	Heavy Oil/ In-Situ Production	LA	1,633	1,870
Northern Peat	ME	Peat/Carbonization	LI	3,065	6,132
Kentucky Tar Sand	KY	Tar Sands/ Solvent Extraction	LA	5,600	16,800
Union Oil Parachute Creek Phase I	CO	Shale/ Unishale B surface retorting followed by upgrading	FA	10,400	N.A.
Union Oil Parachute Creek Phase II	CO	Shale/ Unishale C surface retorting followed by upgrading	LI	42,152	90,000
Cathedral Bluffs	CO	Shale/ Unishale B surface and Modified In-Situ retorting followed by upgrading	LI	15,160	100,000
Seep Ridge	UT	Shale/ In-Situ retorting	LI	1,000	8,000
			TOTALS	115,420	347,340

*FA (Financial Assistance Agreements); LI (Signed Letters of Intent); LA (Authorized, but unsigned Letters of Intent).

Table IX-2

already funded, represented an encouraging array of resource and technology diversity. All together, there were three coal, three oil shale, one peat, and three heavy oil/ tar sands projects.

Just as the Dow PGC was signed, several Board members resigned and the Board lost its statutory quorum. The Corporation was no longer able to transact business. This opened the door for intervention by the administration and Congress as they moved to nominate and confirm new Board members. Further, it was a signal for opponents to the SFC to attempt to brake the accelerating program. These developments will be described in the next chapter—along with a description of how the SFC regrouped and proceeded afresh.

X

Time to Regroup

As discussed in the last chapter, the Corporation was on a roll in the early spring of 1984. Following a slow beginning, the Corporation had approved two assistance agreements and eight Letters of Intent for highly promising projects, and had approved new, targeted solicitations by this time. But times had changed, in energy markets, in the political environment, and within the Corporation itself. This would lead in the course of the year to new legislation, to a new Board of Directors, and to a new strategic direction.

Changing times

Ironically, as the SFC's progress in assembling a portfolio of diverse and promising projects advanced, the economic climate and political support for synthetic fuels was in decline. Energy prices were falling, petroleum imports had declined substantially, and the political enthusiasm for alternatives had waned.

Falling Oil Prices

Falling oil prices particularly contributed to national complacency about energy supply. As shown in Figure X-1, whereas world oil prices had peaked at nearly forty dollars a barrel in 1980, by 1985 they were below thirty dollars a barrel and continuing to fall.

There was even talk of an oil "glut". The declining oil price was inevitably reflected in falling gasoline prices, a reliable shaper of the electorate's mood regarding energy policy. Moreover, falling oil prices and projections of more of the same were significantly

Figure X-1 Oil Prices

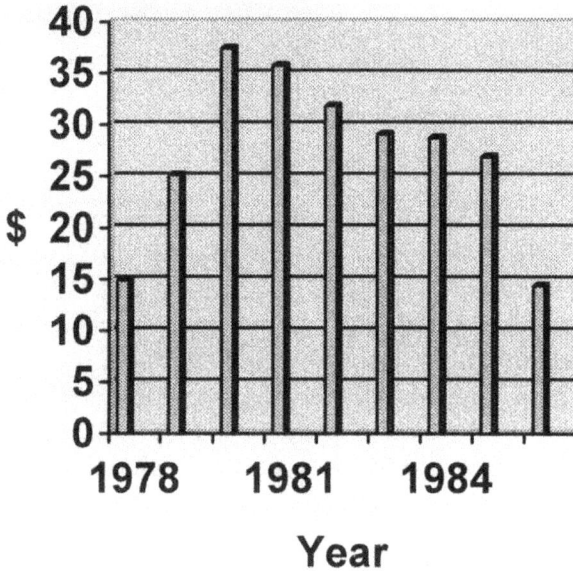

increasing the projected cost of the SFC Program. When oil prices were in the neighborhood of $40 per barrel, it seemed as though relatively little subsidy would actually be expended in the form of the price guarantees. At lower prices, it was clear that the actual outlays would be on the order of billions.

Declining Oil Imports

As seen earlier in Figure I-1, during the year or two preceding the passage of the Energy Security Act, the nation's dependence on foreign oil was increasing despite the implementation of an array of legislation promoting conservation efforts and discouraging the use of oil. By the time of the Iranian revolution in 1979, when gasoline lines appeared for a second time in a decade, U.S. net imports were on the order of eight million barrels a day, up from six million barrels a day at the time of the Arab embargo in 1973. The

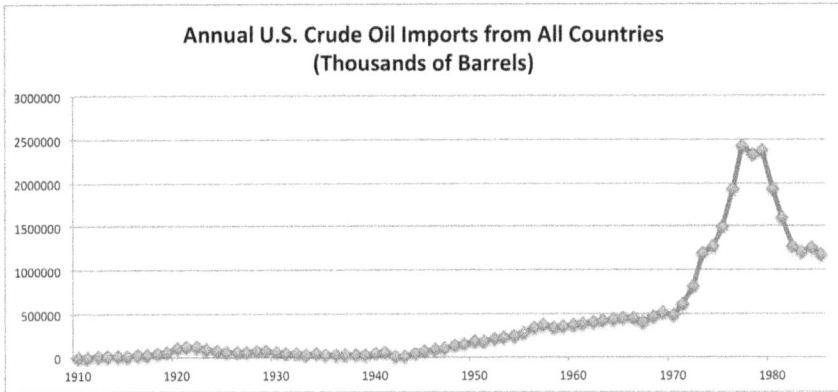

Figure X-2

1979 import levels represented approximately 42 percent of national consumption, despite a vigorous national effort to the contrary. By 1985, however, as shown in Figure X-2, the situation seemed to have changed dramatically.[121] Petroleum imports had fallen to about four and half million barrels a day and constituted only 31 percent of demand.

Many analysts believed that the improvement would not endure. After all, while the 1982 recession, which was the most severe since the Great Depression, had reduced oil consumption along with the GNP, by 1985 the economy was growing rapidly. Moreover, because domestic petroleum reserves and production continued to fall, the dependence on foreign oil would inexorably worsen beyond that experienced in 1979. Various reports concluded that by 1995 import levels would again exceed 8 million bpd and would constitute 48 percent of our domestic consumption (as indeed happened).[122]

But these projections about a situation that would not occur for a decade or more were no longer being taken seriously. And it was clear that sizeable synthetic fuel production would not be in the cards for years to come.

So it made both economic and political sense to cut the SFC program. Yet most observers felt that some continuing program was justified. After all, the United States had experienced an energy crisis in 1973, followed by a quiet period, only to be subject to a new crisis in 1979. Equally importantly, the world remained hostage to oil from the politically volatile Middle East.

It was not clear as to how a course correction would come about, because the SFC funding, for Stage I anyway, was independent of congressional appropriations and the SFC's Board had all the authority it needed to continue the program.

The opportunity for the political process to redirect the SFC mission came with changes in the SFC's Board of Directors.

Events Leading to a New Board of Directors

Five of the seven SFC Board members resigned in the space of a few months in the spring of 1984, so that the Board no longer had a quorum to conduct business. First, Mike Masson's one-year term expired. Subsequently, the president appointed Robert Monks to the Pension Benefit Guaranty Corporation and the lawyers determined that he was not permitted by law to hold both positions. Howard Wilkins, in turn was pressured to leave because of competing responsibilities as a chief fundraiser for Republican candidates for the Senate that year. Political attacks by enemies of the Corporation suggested that as a fundraiser he could not help but eventually be in a position of seeking funds from corporations that were involved in some way with projects that had applied to the SFC for assistance. And, fourth, Victor Thompson, who had been appointed President of the Corporation (in addition to being a Board member) just a month or so earlier, resigned following vicious attacks by a House sub-committee (see Chapter 12) and a charge that one of his private actions could have "appeared" to be a conflict of interest. Because the Board had already lost its quorum, and because his term was to end in a few months anyway, Jack Carter also resigned.

That left only Ed Noble and Vic Schroeder on the Board. The Corporation was stalled and at the mercy of David Stockman at OMB, who had opposed the creation of the Corporation as a congressman and who now vowed to hold up the administration's nominations of new Board members until Congress reduced the funding available to the Corporation by about two-thirds. So the political ball was in Congress's court and it moved to reflect the new energy circumstances.

The Legislative Resolution for 1984

A lengthy and acrimonious debate occurred in Congress between the Corporation's supporters and opponents, which will be summarized in Chapter 13. When the dust settled, Congress had agreed on a reasonable course correction for the SFC. As discussed in earlier chapters, a large synthetic fuels industry did not make economic sense until the technologies were proven at commercial scale (or unless the immediate situation was truly exigent). Also, the SFC experience negotiating with project sponsors was resulting in a de facto smaller program focused on developing a diverse set of technologies. Each project would essentially incorporate one or two units of full-sized technology, but would not attempt a larger operation aiming at achieving full economies of scale. At the time of the congressional debate, the SFC had already incorporated the new thinking in a Basic Business Plan of 1984, which will be described later in this chapter. It was referenced in the congressional debate, and became the essence of the eventual legislative compromise.

Let's look at the legislative actions leading up to that compromise, and then the compromise itself. The legislative steps were as follows:

Early in the summer of 1984, the supporters of the Corporation quickly passed a $2 billion rescission of the SFC's authority hoping that such action would address the Administration's call for a rescission and would avoid the type of debate that eventually took place.

On July 25, in face of the House leadership opposition to having a debate and a vote on rescinding further funds, the opponents of the SFC achieved a relatively rare vote overturning the decision of the Rules Committee so as to permit a debate and vote.

On August 2, Congressman Conte's rescission amendment was debated. Congressman Yates (a supporter of the SFC) offers to amend that Amendment by removing Cathedral Bluffs and Union Phase II (which had letters of intent) from future SFC funding. Congressman Ratchford then proposed amending Conte's proposed amendment with a $5 billion rescission. Both amendments passed the House.

On September 25, the Senate Committee on Appropriations

proposed to rescind $5.2 billion from the Energy Security Reserve and set a formula whereby remaining funds could be used (to be described at the end of this section).

On September 26, Senator McClure modified the Committee language with an Amendment proposed by Senator Metzenbaum to included additional requirements for the SFC regarding the disclosure of information and open meetings.

On October 3, a more restrictive amendment by Senator Bradley was rejected, and McClure offered a compromise, which was agreed to by the Senate.

On October 4, a Joint Resolution of the House and the Senate setting out the compromise as part of continuing appropriations for fiscal year was passed.

On October 10, the Congressional Record recorded the announcement that the Administration supported the Conference Report that accompanied the bill.

The final legislation contained these provisions:

$5.375 billion was rescinded from the Energy Security Reserve (in addition to the $2 billion that had been rescinded earlier in the year).

Of the $8.025 billion that remained available for the original purposes of the Corporation, $5.7 billion was to be applied to projects that had authorized letters of intent from the SFC on or before June 1, 1984 (at this time, eight projects had letters of intent that earmarked a total of $6.8 billion). The language went on to stipulate that if the $5.7 billion were not used for the letters of intent, half of the uncommitted funds could be used to support new projects, but the remaining half would not be available to the Corporation.

The language stipulated that when considering projects for assistance, the Board was to ignore the national synthetic fuels production goals established in section 125 of the Energy Security Act. In effect, this placed paramount emphasis on achieving the diversity goals of the Act.

Additional requirements were placed on the SFC regarding public access to information, conduct of Board meetings in public, and on personal conduct.

While the Conference language did not prohibit the Board from awarding monies to the Cathedral Bluffs and Parachute Creek II projects, it directed the reconstituted Board to review these letters of intent to determine whether the proposed projects would be compatible with the diversity goals of the Act as well as with the suspension of the production goal of the Act. [123]

The language took part of the rescinded monies and made them available to the Clean Coal Technology Reserve.

So, once new appointments were made to the Board to restore the quorum, the SFC would be back in business to reshape its program to accommodate the concerns of Congress and the administration.

The New Board

When the dust settled, Congress had reduced the funding available to the Corporation largely in accord with Stockman's wishes: the SFC's obligation authority was reduced from about $16 billion to about $8 billion. In response, the administration nominated new members for the Board, who were confirmed by the Senate. But the new Board consisted of only five members—Noble, Schroeder, and three others who were sworn in office on December 13, 1984. Because this was an election year and Congress was in recess for the campaigning season, the White House made recess appointments, which are good for a year and do not require Senate confirmation. They did, however, go through confirmation hearings the following spring and were confirmed by the Senate.

The three additions were:

Thomas Corcoran, previously a conservative member of Congress, who was seen as philosophically allied with David Stockman of OMB, who had set off these events of 1984. Moreover, Corcoran had originally opposed the Energy Security Act as being too ambitious, but now saw the merits of a smaller more focused program—which had just been mandated by Congress.

Paul MacAvoy, Dean of the Graduate School of Management at the University of Rochester, had been on President Ford's Council of Economic Advisors and sat on various private Boards of Directors. He had also authored many books on energy and regulatory policy.

Eric Reichl, an eminence in synthetic fuels technologies with experience dating back to the extensive German synthetic fuels effort during the Second World War. He had been the president of Conoco Coal Development Company and was committed to completing a technologically diverse (albeit small) program.

This Board was well equipped with experience on the technical, financial, political, energy, and entrepreneurial fronts. There was, however, a troubling side to the nominations. Once Congress carried out the Administration's wishes, one expected the Administration to fill all the empty seats on the Board to allow the Corporation to function normally. By only nominating three, rather than five individuals, the White House was inhibiting the Board's flexibility. The Board required a vote of four to carry any action. So, with only five members, any two could stall action. Presumably, the White House wanted Corcoran and MacAvoy to be a brake on events. This arrangement had unintended complications that the reader will encounter later in the discussion of the Board's actions with regard to the Great Plains Project.

In any event, the new Board had a full plate of pending business to consider, chief of which was the preparation of a Comprehensive Strategy Report. Not only did Congress require such a report, but its analytic formulation was to play an important role in shaping the Board's priorities in the ensuing year.

THE COMPREHENSIVE STRATEGY REPORT

The Energy Security Act had required the Corporation to submit a Comprehensive Strategy to Congress by 1984, a strategy that was to draw on the experience of Stage I of the program to inform Congress' actions when it appropriated the next $68 billion for Stage II. The prior Board had, however, already reported to Congress that because of the slower than expected pace of funding Stage I projects, the Strategy should be delayed a year. Moreover, it was clearly evident that there was to be no large follow-on program given the shift in energy markets. Nonetheless, Congress needed to be informed of the lessons learned to date, as well as how these would shape the SFC's actions going forward. Consequently, the Board decided to have the staff prepare a strategy *report*, rather than the originally intended strategy. The Board set up a Strategic Planning Committee, which was chaired by Tom Corcoran (who had also been elected Vice Chairman of the Board). Larry Ruff was brought back to the Corporation as a consultant to work directly with this committee[124].

Given the importance of the Report in illuminating the subsequent actions of the Corporation, the remainder of the chapter will be devoted to: (1) outlining the original legislative requirements for a Comprehensive Strategy, (2) summarizing the content of the "Basic Business Plan," which became the blueprint recommended by the Report, and (3) reviewing the benefits that the country would derive from carrying out the Business Plan.

Legislative Requirements for a Comprehensive Strategy

Section 126(b)(2) of the Energy Security Act required that the Corporation submit a Comprehensive Strategy to Congress by June 30, 1984. The SFC did not do so for several reasons. Obviously, it did not have a quorum of the Board that could approve such a strategy. But, more importantly, the prior Board had already concluded that the slow start in getting projects launched meant that the SFC did not yet have the requisite information at hand that would allow the preparation of a meaningful report. Inasmuch as the Act explicitly allowed the SFC to request a delay, the Chairman notified Congress that the Report would not be submitted until June 1985.

In the final analysis, the SFC was not able to submit a strategy to Congress that met the literal requirements of the Act, namely how to achieve production goals of 2 million barrels a day of synthetic fuel by 1992 within an authorization totaling $88 billion. More specifically, it was not feasible to meet the stipulations of Section 126(b)(3) which required that the strategy should:

(D) include a financial or investment strategy prospectus which sets forth the justification for the requested authorization of appropriations;

(E) include findings, based on accompanying comprehensive reports, regarding synthetic fuel projects which received financial assistance prior to submission of such

comprehensive strategy to the Congress with respect to-
(i)the economic and technical feasibility of each such
project including information on product quality, quantity, and cost per unit of production; and

(ii) the environmental effects associated with each such
project as well as projected environmental effects and
water requirements; and

(F) include recommendations based upon accompanying
comprehensive reports concerning the specific mix of
technologies and resource types which the Corporation
proposes to support after approval pursuant to subsection (c) of the comprehensive strategy.

The Act included a procedure for accelerated congressional
consideration of the strategy to be effected by additional appropriations. Be that as it might have been, the changed political circumstances certainly would not have supported appropriations
amounting to another $68 billion. After all, Congress had just cut
$8 billion or so. Moreover, even another $68 billion would not be
sufficient to achieve the production goals of the Act.

The Board, however, felt that Congress was due a comprehensive strategy that laid out where the Board was trying to take the
Corporation in the aftermath of the legislation curtailing the Corporation's resources. Moreover, the Corporation had learned much
in its solicitation and project selection efforts and had a useful story
to tell. The reconstituted Board was anxious to tell it, thereby re-
assuring their supporters and demonstrating that it took the most
recent congressional action to heart.

The Board recognized that while the legislation cut funding and
removed the Production Goals as a factor from the Board's considerations, it left intact the goal of diversity in technology and feedstock. Indeed these factors meshed with the thinking of the new
Board, and made eminent sense with the changing circumstances
of energy markets.

To reflect these factors and to provide a strategically coherent
framework for the planning effort, the Board laid out a Statement of
Objectives and Principles, highlighting the following:

Market forces are keeping oil and gas prices well below the cost of synfuels and were expected to continue doing so for a decade or more.

Unpredictable changes in energy supply and demand could cause world oil prices to increase sharply at any time.

Near-term world oil prices are affected by expectations about future supply, demand, and price.

At any likely energy prices, synfuels production will not be able to expand rapidly and efficiently until some pioneer projects have been built and operated.

Therefore,

Synfuels projects are not likely to be attractive as stand-alone investments in the near future.

A large synfuels industry is not likely to be needed before the end of the century.

There is a national interest in a national synthetic fuels capability that the market alone may not fully accomplish.

The energy outlook, the lead-times involved in synfuels, and the critical role of pioneer plants suggest that some public investment in synfuels is economically justified now.

The objectives that flowed from these conclusions were:

Establish a national synfuels capability that will allow private industry to expand production efficiently and in an environmentally acceptable manner

Improve the national capability to increase production rapidly from selected synfuels options

Improve the national base of knowledge and experience about both the technological and the environmental aspects of selected synfuels options

Assure that commercial-scale processes and projects are designed, built, and operated successfully

Explore a diverse mix of resources, technologies, and geographic regions

Assure that projects continue operating long enough to learn and teach their lessons and that capable people and organizations are developed and maintained.

Building on these principles, the Board adopted a "Phase I Business Plan"[125] to guide them in addressing numerous open issues.

The central function of the Plan was to set out the important priorities for technologies and resources that were to be sought in the projects to be assisted by the Corporation. In parallel with this effort, the staff was directed to prepare analytic support, a set of recommendations for Congress in lieu of the required Comprehensive Strategy, and as much of the detailed information called for by the Act (see above) as possible.

The Phase I Business Plan

The Phase I Business Plan laid out the Corporation's resource and technology priorities in the remainder of the Phase I Program, which are shown in Table X-1.

Several of these technologies were, of course, already represented by funded projects—i.e., the dry bottom, fixed bed gasification by the Lurgi gasifiers in the Great Plains Project, the slurry-feed, entrained-flow gasification by the Texaco gasifiers in the Cool Water Project, and the Union B retort in the Parachute Creek Project. Moreover, the Board explicitly stated that it did not intend to achieve all the priority categories at all costs. Indeed, given the new budgetary constraints, the Statement of Principles and Objectives indicated that each project assisted must make a cost-effective contribution to the Corporation's program by:

Adding significant new knowledge and experience about a nationally important synfuels option, with no more duplication than necessary to provide efficient amounts of insurance and competition;

Assuring that the knowledge and experience generated will be generally available to help subsequent projects by other sponsors;

Being no larger or costlier than necessary;

Having the best possible financial terms for the Corporation, including equitable sharing of profits from favorable events.

Being sound as an industrial undertaking and hence satisfying stringent criteria, including criteria regarding adequacy of the technology data base sufficient to support the design and to reduce the risk of faulty plant operation or adverse environmental impact.

PHASE I PROGRAM PRIORITIES

RESOURCE	PRIMARY PRIORITY	SECONDARY PRIORITY
Coal	Fixed-Bed Gasification - Dry Bottom - Slagging Entrained-Flow Gasification - Slurry Feed - Dry Feed Fluid-Bed Gasification	Direct Hydrogenation Other Resources and Technologies
Shale	Surface Retorting of Western Shale - Union B Retort - Other Retorts Modified In-Situ Retorting of Western Shale	True In-Situ Surface Retorting of Eastern Shale
Heavy Oil/ Tar Sands	In-Situ Recovery Mining and Surface Processing	

Table X-1

The Report examined each of the categories established under the Business Plan and identified the progress that had been made toward realizing an actual project in terms of: projects that had received assistance from DOE or the Corporation, letters of intent that were still pending, projects that had applied under solicitations that were still being considered, and projects being sought by the Corporation under the solicitations that the newly constituted Board had just issued (see the discussion of Targeted Solicitations in the prior chapter regarding eastern coal gasification projects and projects involving mining and surface processing of tar sands).The Report estimated that the Business Plan could be implemented within the appropriation authority left to the Corporation. The Report summarized the cost estimate as shown in Table X-2 below.

The Report also noted that the funds would only be outlaid over a considerable number of years. In 1985 dollars, the dis-

ESTIMATED COST OF IMPLEMENTING THE PHASE 1 BUSINESS PLAN

RESOURCE CATEGORY	OBLIGATIONAL AUTHORITY	
New Commitments	Low	High
Priority coal (including Great Plains)	$3 billion	$4 billion
Priority Oil Shale	$2.3 billion	$3 billion
Priority Tar Sand/Heavy Oil	$200 million	$500 million
Secondary Priorities	0	$800 million
Total New Commitment	$5.4 billion	$7.8 billion
Existing Commitments		
Cool Water		$120 million
Dow Syngas		$620 million
Parachute Creek		$400 million
Total Existing		$1.14 billion
TOTAL OBLIGATIONAL AUTHORITY	$6.54 billion to $8.94 billion	

Table X-2[6]

counted present value of the high authority would have been approximately $3 billion. In other words, the final expenditures for the commercialization program represented by the Business Plan would be less than had been expended by the federal government on synthetic fuels R&D over the prior couple of decades. This was a far cry from the ambitious goals set out by the Energy Security Act, and it was eminently in keeping with the intent of the legislation that had been passed the preceding year.

Benefits Anticipated From Accomplishing the Plan

The Board made it clear that the Business Plan was fully responsive to the legislative mandate to achieve diversity of technology and resource in the projects supported by the Corporation. Of course, it had to stay within the confines of the reduced obligational authority left to it. But one could still pose the question: does the program still make sense in terms of national needs and changing energy circumstances? The Board looked at this question in depth, and the Report provided the analytic underpinnings of the conclusions that the Business Plan did, indeed, make sense in both economic and strategic terms. And what might such terms be? First, would the Program provide strategic capability in the event of a major energy disruption? Second, would the Program make economic sense, if there were no disruption, and the country simply had to contend with a natural orderly evolution of energy markets? The Board concluded as follows.

As often voiced by Chairman Noble, the central rationale for the SFC Program was that it provided the nation with an "insurance policy" in the event of a major energy dislocation. While this concept was not expanded upon at length in the Report, it was acknowledged as almost self-evident. This concept was, in its essence, that if the Program were a success so that the nation had available to it an array of proven and most economic synthetic fuels alternatives, then private industry could rapidly expand production from those alternatives when circumstances so dictated.

It would, however, take industry on the order of three years to expand capacity to any degree since the output of the proposed Phase I plants would be negligible (perhaps on the order of 100,000

BOED or so). If so, was the "policy" sufficient for the potential circumstances? Was it even worth doing? What was the nation actually buying?

Essentially, it would have been buying time (and all the accompanying economic benefits under those circumstances), it would
have been buying the knowledge of which were the best and most
economic processes with which to build an expanded industry, and
it would help avoid the likely pitfalls of investing in flawed large-
scale technologies under exigent circumstances.

To see whether this was an economically sufficient rationale,
one had to quantify hypothetical circumstances and outcomes.
Surely, if the insurance policy was too expensive in light of the payoff, we would probably be better off just hoping for the best.

This assessment was, of course, not much different than the
one faced by South Africa over the last several decades. In the
mid-1950s, that nation had decided to build coal liquefaction facilities using German war technology to draw on their sizable coal
resources so as to avoid the potential consequences of an oil embargo. They obviously were prescient on this score. As discussed
elsewhere, they built a synthetic fuels industry called SASOL in
three phases (the late 1950s, after the embargo of 1973, and after
the Iranian revolution of 1979). When these were completed, that
nation produced roughly half of all their liquid fuels needs from
coal-derived liquids.

The situation that the United States had to plan for was not,
however, one likely to be as exigent. The South Africans had to plan
for a tight embargo. At worst, the United States would have only to
worry about a multi-nation embargo (perhaps the Arab states). But
there would always be oil-producing nations willing to sell to the
United States, and since oil is virtually a fungible commodity under such circumstances, the practical situation would be more one
of relative scarcity and sharply higher prices for the entire world.
Given that scenario, and assuming that an embargo would be one
of a long duration, the question facing the SFC planners was how
would synthetic fuels play a role as formulated under the Phase I
Program? Remember also, that the nation had been building a strategic stockpile of oil to help ameliorate the effects of oil shortfalls. In
effect, the situation would likely be one of sharply higher oil prices

that would fall as more and more synthetic fuel production (and conservation efforts and other alternative fuels) took effect.

To determine the net benefits of such a course of action, the Corporation's staff modeled the financial outcomes resulting from having the Phase I Program and not having it under the above scenario. The chief differences between the two hypothetical outcomes were that: without the program, full-scale plants would require close to ten years to design, build, and eliminate operating problems and with the program, only three years would be needed. Moreover, without the program, avoidable expensive mistakes would be made. This latter consequence was not quantified. But just being able to build the plants more rapidly was estimated to have benefits on the order of $14 billion to $18 billion (in 1985 dollars).

This level of savings would not justify the full program set out under the Energy Security Act, but would support the concept underlying the Phase I Plan—if such events occurred. This logic and calculation justified insurance programs across the board. But that outcome was not certain. Could the $3 billion cost of the proposed program also be justified or at least ameliorated if the worst-case scenario never occurred?

Extending the analysis above, the Board found that the program would provide substantial national benefits even under circumstances associated with relatively stable evolution of energy markets. The economic benefits would be derived from two principal sources: from reducing the cost of synthetic fuels so that they would be available to energy markets sooner than would otherwise be the case, and from their indirect effect on the calculations of oil producers that would slow the rise in energy prices.

The first argument is the more conventional and straightforward. It rests on the calculation that such intervention by the government in energy markets would help the nation to capture economic benefits that would not accrue simply by the private sector responding to market forces. How might this occur?

In essence, the argument relies on the experience that the risks and uncertainties inherent in pioneer plants inevitably result in their products costing more—significantly more—than would the products of follow-on plants. For many patentable products, these higher costs are tolerable because competing products (if any) are inferior even at the high cost, and because during the subsequent period of a quasi-monopoly, the developer can recoup his investment.

This model obviously does not obtain for energy markets. Synthetic fuels hold no advantage (or disadvantage) over natural petroleum or gas and thus cannot be sold at a premium. Moreover, while the new technologies are patentable to be sure, the developers will likely only receive relatively modest royalties. They will not be able to own and control the new production facilities.

Now, why are the costs of pioneer plants inevitably higher than those of the next generation? There are four major contributing factors. The most important of these stem from the uncertainties confronting new technologies. Most new plants suffer from cost overruns that occur from dealing with unknown circumstances. They take longer to complete, and they take longer to achieve design rates of production. Let's turn to the South African SASOL example again. Despite the fact that "proven" German technologies were being used, the larger scale and transplanted circumstances introduced major unexpected problems that profoundly affected the timing of full production. After the formal end of construction, the SASOL I facility did not achieve full design production levels for several years. SASOL II, on the other hand achieved design levels within about a year following the end of construction. SASOL III did better still. Such differences in timing have, of course, profound effects on the rate of return seen by the investor. That is why nobody likes to be first.

Or, to look at the other side of the coin, to achieve the same rate of return, the selling price of the product would have to be significantly higher for the first-of-a-kind plant as compared to the follow-on.

A second factor leading to cost reductions is simply the learning that occurs in building and operating a major facility. Often, one can only learn by doing. SASOL I continued to experience improved economics 15 years after the end of construction. Systems are redesigned, bottlenecks are eliminated, and operating economies are found. And, follow-on plants benefit from insights regarding basic design. All of these effects are often collectively referred to as "learning benefits."

A third factor considered in the SFC analysis was the rate of return that would be required by the investor (or the lender if willing to lend at all without guarantees) would be higher to compensate for all of the uncertainties. Sponsors of projects applying to the Corporation indicated that they often looked to rates of return of the order of 30 percent for developmental facilities (usually based on projected costs, rather than actual). Because the SFC was effectively taking over the key completion and market risks through the loan and price guarantee mechanisms, supported projects tended to look for rates of return on the order of twenty percent. Follow-on, risk free facilities, however, might well be content with returns in the middle teens.

A final factor to be considered is that of economies-of-scale of a facility. Strictly speaking, this factor need not be different between the pioneer and follow-on facilities, since they could all be built at the same scale. Indeed, many of the projects that had applied to the Corporation early on did contemplate facilities that would attempt to capture full economies of scale in production. However, as a practical matter, it usually makes greater sense to build only the basic technology at full scale, and then replicate multiple "trains of components" for fuller economies when the technology is perfected.

The SFC staff attempted to estimate all of the above affects for each of the principal resource areas being supported by the Corporation. The summary results are presented in Table X-3 below.

Since these results are critical to the argument in favor of the Phase I Program, let's look at some of the key assumptions that went into this analysis. A great deal of effort went into establishing the appropriate capital costs to be used. In that regard, the Corporation benefited from having the most complete set of cost data on synthetic fuels plants available anywhere. Yet, the analysts needed to treat the input with appropriate care. As noted earlier, no one really knows what a pioneer, large-scale plant will cost until it is actually built and operating. At the outset, cost figures tend to be clouded by the optimism of proponents and to be too free of the probable negative effects of the unknowns. Throughout this analysis, Len Axelrod, Vice President for Technology and Engineering, cautioned the staff not to take the numbers further than they de-

SUMMARY OF PIONEER AND FOLLOW-ON PLANT OUTPUT COSTS

	Pioneer Plant	Second Generation	Third Generation
COAL			
Plant Scale *BOED*	7,500	20,000	40,000
Unassisted Levelized *Cost ($ per BOED)*	125	40-50	25-40
OIL SHALE			
Plant Scale *BOED*	14,500	30,000	45,000
Unassisted Levelized *Cost ($ per BOED)*	110	45-50	35-40
TAR SAND			
Plant Scale *BOED*	5,500	10,000	40,000
Unassisted Levelized *Cost ($ per BOED)*	90	55-60	40-50

Table X-3[127]

served. And he was probably still somewhat uneasy with the apparent concreteness of the final results.

Nonetheless, we knew a great deal more than would have been the case, absent the SFC Program. The Corporation's insistence on projects having a relatively advanced level of design meant that most of the cost estimates to be used in the analysis had much smaller uncertainties than would otherwise have been the case. Also, at that point the Great Plains, Cool Water, and Parachute Creek facilities had been constructed.

Moreover, because of multiple submittals of proposals for the same projects, the data available to the Corporation, often involved estimates for projects of various sizes using the same technologies. So, while there were still uncertainties in each estimate, the ability to separate out the effects of economies-of-scale existed. Another issue to be dealt with was how to treat the proprietary nature of the cost data held in confidence by the SFC. Since the point of the analysis was simply to quantify effects and trends, the numbers did not have to be project specific. Accordingly, the staff developed composite cost numbers using a few estimates for advanced designs in similar technologies. Thus, for example, one could not mix oil shale and coal technologies. Nor, would it have been proper to mix coal gasification and coal liquefaction technologies. So, for example, the above numbers for coal plants only include cost data for projects producing medium-Btu gas.

With the above approach, the data in hand seemed to be sufficient to establish the starting costs of pioneer plants. The next analytic hurdle was how to estimate the benefits of "learning" in terms of the decreasing costs of operating an existing facility over time, and in estimating the decreasing costs of follow on plants (exclusive of the effects of increasing scale). Here, the staff decided to draw on extensive experience gained in the expansion of similar industries that had been based on new technologies.

To this end, the analysis drew on work performed by Hagler, Bailly and Co.[128] This Report examined the experience of three industries having significant relevance for synthetic fuels i.e. low density polyethylene (LDPE), ammonia, and alumina refining industries. Indeed, ammonia is itself a synthetic fuel under the definition of the Act. But the specific aspects of these industries that make their experience pertinent include a large solids-handling component, chemical synthesis, and high capital intensity. The pattern of product cost reduction over time for the three industries was remarkably consistent. Each experienced a learning rate of approximately 80 percent as summarized in Appendix D.

The SFC analysis did not claim that synthetic fuels would necessarily derive such dramatic cost improvements. It did argue, however, that cost improvements largely came from actually building and operating plants, with the new generation facilities embodying the learning gained from the earlier facilities. More specifically, the analysis assumed that second and third generation plants would be scheduled at an optimum pattern that would allow predecessor plants to have come on line with several years of experience to shake out bugs, before the next plants were designed and built. For example, second generation plants were not to come on line until the pioneer plants had operated for more than five years. With these assumptions and an approach performed by Rand,[129] the levelized costs of the second generation plants were assumed to be approximately 8 to 22 percent lower. These cost reductions were assumed to be derived almost solely from operating experience.

For the third generation plants, however, which were not to begin design until three years after the start of design for the second generation, there was to be full capture of construction learning cost reductions (but not operational learning since the second generation plants would not have yet come on line). These assumptions resulted in a reduction of levelized costs for the third generation plants of about 12 percent.

The specific assumptions for reflecting the expected effects of economies-of-scale were substantial as well. Cost reductions of approximately 5 to 30 percent were assumed for second-generation plants and of approximately 0 to 20 percent for third generation plants. (The spread in the estimates reflected the range of cost estimates associated with the actual projects evaluated by the Corporation.)

Finally, with regard to the above Table, the assumptions used to reflect lower financing costs that would result from reduced risk and uncertainty were substantial as well. Using assumptions believed to be typical of industry, pioneer plants were expected to be financed 100 percent by equity having a 21 percent real after-tax rate of return on equity. Follow-on plants were to be financed 50 percent with debt (increasing the leverage) having an interest rate equal to 120 percent of the 20-year Treasury bond rate, and equity receiving only a 17 percent real rate of return. These assumptions resulted in calculated cost reductions of approximately 5 to 30 percent for second generation plants and 0 to ten percent for third generation plants (depending on how capital intensive a project is).

Two important findings emerge from Table X-3. First, no one

would invest in a pioneer synthetic fuels facility—in our lifetimes anyway—on the basis of expecting a profit. Energy prices simply will not reach sufficient levels on a sustained basis. On the other hand, once the technologies are proven, energy prices in constant dollars would not have to get any higher than they did in 1980 for the private sector to be able to expand a synthetic fuels industry on a profitable basis. In other words, the existence of proven technologies effectively would establish a cap for energy prices.

This finding, however, requires an important caveat. Might not a firm take a loss on a pioneer plant so as to be in a position to capitalize on the rights to the technology when it becomes profitable? Possibly so. After all, other industries as we have noted have done so. Similarly, the private sector has invested in a number of substantial pilot facilities that have produced no immediate return. But, as a practical matter, full commercial-scale facilities are so expensive--a minimum of half a billion dollars (1985 dollars) for most shale and coal technologies--that no firm would realistically take such a risk.

A second finding explains the escalating cost phenomenon that has bedeviled synthetic fuels for the last few decades. Namely, over this period, proponents of synthetic fuels regularly estimated what the market price for competing fuels would have to be for a plant to be economic. Then, when the price reached that level (e.g., after each price spike in the 1970s), the proponents said that the prices needed to get higher still. The explanation for this pattern seems to be apparent from the Table X-3. The proponents are probably citing costs for a relatively mature technology as opposed to what market prices would have to be to justify going ahead with the pioneer facility.

The central point of this section is that undertaking the Phase I Program would have contributed a positive return to the nation even without international disruptions in the energy markets. That argument rests on projections of energy prices that suggested that second and third generation plants would be profitable by the turn of the century. If so, and if the technologies were proven, then synthetic fuels would be cheaper than imported oil, and the nation would reap positive returns commensurate with the number of plants actually built. The SFC analysis projected plant construction that might occur consistent with the price projections in order to estimate the value of the program to the country. The net return was on the order of $3 billion to $5 billion using a median price scenario. This return was net of the cost of the Phase I Program.

Potential Benefits from Lower World Oil Prices

Another major—although more abstract—benefit that could accrue to the nation from the program, even in the absence of energy disruptions, would be that occurring from a slowing in the rise of world oil prices by virtue of the existence of proven synthetic fuels technologies. In deriving the potential value of this effect, the Corporation's staff drew on the theory of the Hotelling Rule, named in honor of Harold Hotelling, who first advanced the theory in 1931.[130]

The gist of this theory describing the pricing of exhaustible resources is that long before the resource, such as oil, is physically depleted, its owners foresee the time that it will become scarce and dramatically increase in value. Thereupon, the price of the resource begins to be affected by the owners' expectations of future prices, as much as by current production costs. If the price is expected to rise relatively rapidly, many owners will withhold supplies from the market in expectation of receiving still higher prices. Conversely, if prices were not expected to rise rapidly, owners would tend to produce and sell more currently than later and invest the added proceeds elsewhere. Market equilibrium would occur when owners expect prices to rise just fast enough that leaving their resource in the ground would be expected to pay off as well as, but no better than, selling it and investing the proceeds.

For a "pure" exhaustible resource that is subject to high market demand but exists in low-cost, known, and large but non-reproducible deposits, this theory suggests that its market price will increase at the prevailing rate of interest. Petroleum is considered to be similar to such a resource. Coal and shale, on the other hand are so vast in amounts that their prices are likely to increase at a slower pace related to production costs.

In applying this theory to the Phase I Program, the analysis assumed that the effective cap that synthetic fuels alternatives would place on oil prices would affect the behavior of the resource owners such that prices would escalate less rapidly than they otherwise would. The theory (as understood anyway) was not quantitative enough to predict the price effect. Nonetheless, the staff attempted to assess the potential of the effect. Specifically, they assumed that if the effect were to keep oil prices $0.25 a barrel less than they otherwise would have been (and increase proportionately with the future price of oil), then the nation would save from $450 million to $1.3 billion a year (increasing with time). Counting only the savings on imports to the United States, the discounted present value (at a ten percent rate) would be on the order of $8 billion. This benefit would be in addition to the amount attributed to the savings from having the technologies on hand when the prices justified replicating the plants (i.e. the price scenarios for the earlier benefit analyses assumed the Hotelling effect).[131]

In summary, the combined benefits from the program under stable energy circumstances could be in the range of $11 billion to $13 billion—or a net $8 billion to $11 billion taking into account the cost of the Program.

The Board believed that it had produced an explicit, analytically sound plan to carry it through Phase I of the Corporation. It made economic sense, met strategic needs, and seemed to be in tune with the new political realities. So, the Board was primed for rapid action to carry out the Business Plan as it addressed an array of pending business—the subject of the next chapter.

XI

The Last Accomplishments

The Board, having formulated the Comprehensive Strategy Report outlined in the prior chapter, had a clear picture of what it wanted to accomplish in Phase I with the obligational authority remaining to the Corporation. The Board at this time had a full plate of unfinished business to undertake. As the reader may recall from Chapter 9, the first Noble Board had authorized eight letters of intent and had approved three solicitations under which the Corporation had received proposals during the interregnum. The letters of intent could, of course, not be converted into approved commitments absent a quorum; nor could action be taken on the new proposals. In addition, the Board needed to launch whatever new solicitations would be necessary to complete the Program envisioned by the Business Plan.

All this amounted to an ambitious effort on the part of the Board—one that kept it meeting frequently in the year following its confirmation in December 1984. Let's review these accomplishments.

THE EIGHT LETTERS OF INTENT

The changed legislative mandate required that the structure of the letters of intent be revisited. Despite the extensive effort on the part of the Corporation as well as of project sponsors that went into reviewing proposals and negotiating the terms of potential assistance, the implicit "deal" might be undone. True, Chairman Noble signed the letters indicating that he would push for Board approval of assistance contracts if the project sponsors carried out actions to fulfill the conditions contained in the letters. But, with the action of Congress to curtail the obligational authority available to the Corporation and to modify its mission, the Corporation was placed in a position vis-à-vis the sponsors akin to "that was then, this is now."

How to proceed? The Board wished to move expeditiously such that it could consider several projects at each meeting early in the year. To lay the groundwork for potential Board action, it requested the staff to undertake a new *programmatic* review of the eight projects and recommend whether the projects were still appropriate in light of changed legislation and changed policy. To that end, "B" teams were set up to reevaluate the projects. Individuals making up the B teams were different from the original "A" teams. Otherwise, the same evaluation expertise was brought to bear. (During the time that had elapsed since the original Board actions, circumstances had changed for some of the sponsors as well. For example, Texas Gas was acquired by the CSX Corporation, and the new management asked that the Kentucky Tar Sand Project be removed from further consideration.)

Once the reevaluations were complete, the Board considered whether it wished to accept the terms of the letters of intent or whether the Corporation's altered mission required new terms to be negotiated.

The following sections summarize the results of the reevaluations, as well as describe the Board's actions regarding each of the projects. Two of the eight projects (Parachute Creek and Forest Hill) received awards of assistance, and two (Great Plains and Seep Ridge) came very close; these four projects' re-evaluation and renegotiation of assistance will be described most fully.

Parachute Creek - The Augmentation Program

Unocal and the Corporation had signed a letter of intent that would have awarded the Parachute Creek Project another $2.7 billion of assistance to build a Phase II facility having a production capacity of 42,000 BOED. Unocal had already built the Phase I facility with the incentive of $400 million of price guarantee assistance.

The Phase II Project had been extremely attractive to the Corporation on a variety of grounds. But some of them were no longer valid. First, the Phase II Project was to employ a Union C retort rather than the Union B, which was in place in the Phase I facility. While both would use the basic "rock pump" retort, the "C" retort would include substantial technical advances, which Unocal did not include in the "B" retort so as to control the amount of risk being assumed. These advances would increase the technology's energy efficiency by capturing the energy remaining in the carbon on the spent shale and by capturing the heat in the hot shale emerging from the top of the retort. If proven out, the "C" retort would be the one used in the commercial expansion of a larger shale oil industry.

Secondly, the project would be a cost effective contributor to the production goal of the Energy Security Act. At 42,000 BOED, it was by far the largest project the Corporation considered. Because it was large enough to virtually achieve the economies-of-scale inherent in the technology, the amount of assistance per barrel produced was much less than for the projects represented by the other letters of intent. For example, the Dow Syngas Project received a price guarantee commitment with $620 million of obligational authority for a facility having a capacity of about 5,170 BOED. Per barrel of output, the Parachute Creek Phase II assistance would have been half that of Dow.

These potential project strengths notwithstanding, two crucial conditions had changed since the signing of the letter of intent. Most notably, when Congress cut the obligational authority available to the Corporation in half, it made it infeasible for the Corporation to devote $2.7 billion to a single project. In keeping with the reduced obligational authority and the continued interest in diversity of technology, Congress specifically eliminated the production goal from having a role in project selection—in effect, signaling a desire for multiple smaller projects.

Equally importantly, Unocal had still been unable to get the "B" retort of the Phase I facility to operate at design levels: while the project had completed construction by the end of 1983, by 1985 it was still not reliably operable. Clearly, neither Unocal nor the Corporation was about to sign a binding $2.7 billion commitment based on a technology experiencing such difficulties.

The "B" team reviews endorsed the programmatic value of building a "C" retort, and the Business Plan explicitly identified the new retort as an objective of the Corporation's program. But there would have to be a sharply smaller commitment. The Corporation's staff and Unocal reviewed options into the summer of 1985, and concluded that an "Augmentation Program" was the most sensible way to proceed. In essence, the Augmentation would involve adding a fluidized-bed combustor (FBC) to the "B" retort of the Phase I plant along with associated heat recovery equipment, in effect creating a "C" retort. While this would entail a complicated process with overall higher costs that would be required if a "C" retort were built from scratch, the incremental obligational authority needed would be far less than would be needed for an entirely new facility.

With this concept in mind, the staffs skipped a letter of intent and began to develop directly the language of an "Amended Commitment," resulting in the following changes to the letter of intent agreement.

The Amendment contained a key condition, however. Unocal was to develop the engineering design and detailed cost estimate for the Augmentation Effort by a date certain. This effort was likely to cost about $10 million or so. If the estimate exceeded $286 million, then Unocal was not obliged to proceed with the Augmentation Program, and the incremental $500 million would be de-obligated.

In addition, a new price guarantee formula was to be melded with that of the original assistance agreement. The original price guarantee was $42.50 per barrel (adjusted for inflation) for diesel and jet fuel. Those market prices to be determined by the Depart-

	Original Terms	New Terms
Plant Capacity:	42,000 BOED	10,000 BOED
Maximum SFC Obligation:	$2.7 billion	$500 million
Amount of Loan Guarantee:	None	$327 million
Amount of Price Guarantee	$1.55 billion (B retort)	$173 million annually
	$1.95 billion (C retort)	Up to $500 million after loan repaid
Guaranteed Price*:	$60.00/barrel (B Retort)	$67.87/barrel
	$67/barrel (C Retort)	

Adjusted for Inflation.

Table XI-1[132]

ment of Defense, which had the ongoing option of purchasing the direct output of the facility or of taking the equivalent in conventional fuels supplied by Unocal elsewhere. The Amendment converted the original fuel reference to upgraded shale oil instead of the defense fuels. Since the upgraded shale oil, while superior to crude, was not as valuable as diesel and jet fuels, the guarantee price was reduced to $37.87 (adjusted for inflation).[133]

The project would receive only this level of price support until the fluidized-bed combustor was completed. Thereupon, the project would receive a price support supplement of $30 per barrel (adjusted for inflation). Unocal would be eligible to receive price guarantee payments for a period of ten years from the date of receiving the first price guarantee payment (or until the maximum amount of obligational authority was reached). Unocal would be liable to make profit-sharing payments for a total period of 16 years.

The Amendment Agreement was signed on October 16, 1985.[134] The Board approved the terms and conditions implementing the guaranteed loan (once Unocal had negotiated these with the bank making the loan) on January 21, 1986.

The Forest Hill Project

Since Chapter 9 already described the Forest Hill Project and its technology, the following section merely describes the events that led to the final assistance agreement.

The initial step in these events, of course, was the "B" team review instituted by the Board. In light of the priorities of the Business Plan, which called for a heavy oil/tar sand project utilizing an in situ technology, the project easily cleared the programmatic hurdle. Also, given its modest claims on the remaining obligation authority, it posed no hard choices for the Board. The chief effort was to determine whether the terms of the letter of intent, which had been negotiated over a year and a half earlier, were still sufficient to make the project viable.

In the event, the terms of the financial assistance embodied in the letter had to be modified only slightly. For example, the guaranteed loan authority was increased to $24.4 million (plus a $2 million allowance for accrued interest not paid in the event of a default) from the $22.57 million contained in the letter of intent. The total obligational authority was increased commensurately, but the price guarantee level of $40 per barrel of the initial 500,000 barrels of production and $37.50 per barrel thereafter was not changed. Accordingly, converting the financial terms of the letter of intent into the language of a final contract was straightforward.

The drafting team did undertake considerable effort with regard to due diligence in ensuring that the contract would be a prudent one in light of the Corporation's mission, as well as a thorough analysis to demonstrate that this heavy oil project was eligible under the definition of eligibility in the Act. (Recall that the Act only made eligible those heavy oil projects that would not be economic without government assistance and which entailed technical risks comparable to other synthetic fuels projects.) Also, the team engaged consultants to verify the amount of oil likely to be recoverable from the reservoir with the oxygen fireflood technology, and the readiness of that technology for such application. In the end, the staff's report was positive, and the Commitment to Guarantee was approved by the Board and signed by all parties on September 25, 1985.[135] It was the third award made by the Corporation.

The Great Plains Project

The Corporation's staff probably devoted more time to completing a Commitment with Great Plains than it did for any other project. In the end, political reasons made it fruitless. The highlights of what turned out to be an unnerving process in some ways signaled the end of the Corporation.

The process began easily enough working from the letter of intent that had been signed on April 26, 1984.[136] The Business Plan welcomed the Great Plains technology as a high priority and the "B" team had no difficulty in passing the project over the programmatic hurdles. Further, the negotiating team had little difficulty in reaching agreement with the project's sponsors on the amounts of assistance and the terms of the final Commitment. The approach entailed the staff negotiating two different sets of terms *and* going through the extensive documentation needed for a final Commitment—beyond the original terms that were negotiated as part of the

	Original Terms	New Terms
Maximum SFC Obligation	$790 million	$820 million
Guaranteed Price*:	$10.00/MMBtu (3 years)	$10.00/MMBtu (until 3/31/88)
	$7.50 MMBtu (thereafter)	$6.50/MMBtu (thereafter)

*Adjusted for inflation.

Table XI-2

letter of intent. The first of these negotiated terms were presented to the Board at its meeting on April 22, 1985. At that time, the board received the staff's analysis that confirmed that the project met the criteria of the Business Plan. In addition, the Board reviewed the key financial terms being discussed.

The essence of the proposed deal was that the sponsors would be required to commit an additional $190 million of equity to the project, which together with all the positive cash flow until March 31, 1988, would be applied to prepayment of the DOE-guaranteed loan. In addition, the rights of the sponsors to abandon the project during the initial reinvestment period were to be made substantially more restrictive than was the case under the existing agreement with DOE. In exchange, the Corporation agreed to the modified assistance terms explained in Table XI-2.

As can be seen, these provisions were not all that different from those contained in the original letter of intent. In exchange for a modest increase in maximum SFC obligation, the sponsors would have committed an additional $100 million of equity and would have applied all tax benefits and profits to the prepayment of DOE-guaranteed debt for approximately three years after the closing of the commitment. This agreement reflected the passage of a year or so, particularly changing energy prices.

At this point, in my opinion, the Board overtly entered the political arena, abdicated at least part of their responsibilities, and added to the pressures building for the demise of the Corporation. The Board had all the authority it required from the Energy Security Act to approve the proposed price guarantee commitment with Great Plains. But, for whatever reason, they requested further advice and analysis on the alternatives from DOE, the Treasury Department, and the Office of Management and Budget. At that Board meeting,

Jan Mares, Assistant Secretary of Energy for International Affairs and Energy Emergencies, appeared on behalf of Secretary Herrington and endorsed the understanding that the staff had reached with the project. Further, he urged that the Board move quickly to conclude an Agreement.

So, what was the Board up to? Its action probably grew out of the fact that DOE was deeply involved in the project through the guaranteed loan, and, perhaps, out of a concern that the government might be perceived as providing the project with two subsidies. Strictly speaking, while there would be two commitments, the only subsidy would be the price guarantees being provided by the Corporation. Unless the project defaulted on the loan, which they would not because of the price guarantees, no funds would ever have been expended by the government as a result of the loan guarantee.

This was a classic situation of not asking a question unless you were certain that you would like the answer. The DOE response involved some vague concern that the structure of the deal did not adequately provide for the long life of the project, during which time the nation could be assured of achieving full learning benefits from the project. As far as the staff could discern, this concern came from nowhere and was ominous. The price guarantees would have enabled the project to operate for a good many years—ample to gain the learning benefits, and certainly on the same order of time that would have been the case for the other projects receiving assistance from the Corporation. Moreover, the project would operate much longer with the price guarantee assistance than it would without. Given Herrington's eventual position, one could in retrospect see that the DOE concern was tactically motivated and had no substantive basis whatsoever. Yet Secretary Herrington's letter of May 21, 1985, was otherwise supportive. It stated, "We are not opposed to a price support agreement. ... we believe every effort should be made in negotiating such an agreement to ensure long-term operation of the plant ... . In conclusion, we urge the Board of Directors of SFC to approve and execute an appropriate price guarantee assistance agreement with the sponsors of GPGA as soon as reasonably possible."[137]

Chairman Noble's response after the Board meeting accepted the Secretary's premise and included the following language:

> They [the Board] also agreed that the optimum assistance agreement would ensure long term operation of the plant and make maximum additions to the synthetic fuels data base.
>
> Further, the members felt that the most positive way of obtaining long term operation of the Great Plains project was to restructure its debt repayment schedule. The schedule for payments on the DOE guaranteed debt to the Federal Financing Bank (FFB) must be extended. This would be consistent with successful operation of a first generation large-scale synfuels plant. Because of the heavy debt, proposed SFC price guarantees have had to be extremely large during the initial operating period of the plant. The debt interest and repayment schedule should be restructured to remove this impediment. Repayment of interest and principal must be postponed until the later years of plant operation when net operating income is likely to be available.
>
> Rescheduling of payments to the FFB would allow project operations to go forward based on price guarantees that may be considerably reduced from those now in the proposed SFC agreement. Lower price guarantee levels would extend SFC support over a longer period, resulting in lengthy operation of the plant. Thus, to achieve the goals set out in your letter of May 21, the Administration, principally DOE and FFB, should provide a realistic schedule of debt interest and principal repayment.
>
> The SFC is prepared to sustain a high level of negotiating activity over the next few weeks to produce "an appropriate price guarantee assistance agreement with the sponsors of GPGA." This will necessarily require daily interaction with those who will be undertaking to restructure the FFB payments. Therefore, I propose that you set up an ad hoc interagency task force of senior officials in the appropriate agencies in which the Corporation may participate.

The Corporation staff did proceed to renegotiate once again the terms of the deal along the above lines, and did revise the proposed contractual documentation. That revised agreement in principle would have had DOE use its $673 million default reserve to reduce the existing project debt of $1.54 billion with the FFB. The project in turn would have issued the DOE a note providing for repayment of the $673 million in accordance with the project's financial performance, but in any event by lump sum payment in 2009. Then, the SFC would pay up to $720 million in price guarantee assistance over a ten-year period. These price guarantees, together with the rescheduling of the debt, would have provided the reasonable likelihood of long term operation. The requirement for the project to make profit-sharing payments was also preserved, but for a 25-year period.

Thus, all the points in Secretary Herrington's letter were addressed. But for naught. On July 29, 1985, he informed the Corporation that the DOE would not cooperate in the restructuring of the loan, which the Directors had required as a precondition to an award. Consequently, the Board took no further action on the proposed award to the project.

Then, without the prospect of receiving any price guarantee assistance, the project's sponsors defaulted under the terms of their loan guarantee agreement with the DOE, as they had indicated they would when they had applied to the Corporation for assistance. And, the government had to make good on the one and a half billion dollars of defaulted loans. So the government was financially worse off than it would have been if the price guarantee agreement had been consummated.

The Seep Ridge Project

The Seep Ridge Project came very close to being the fifth project to receive an assistance award from the Corporation. Very close. But the normal changes in the business arrangement and the fall in energy prices since the letter of intent had been negotiated, combined with an unexpected legal decision regarding Indian Affairs, made it miss out by a matter of days.

Chapter 9 described the project and the terms of the letter of intent, which had been authorized on December 1, 1983, and signed on June 22, 1984. The project was to be located in Uintah County, Utah, and was to produce 1,100 barrels per day of raw shale oil using the fully in situ Geokinetics "LOFRECO" process. The terms included loan and price guarantee assistance totaling $45 million.

The "B" teams completed their review of the programmatic merits of the project in light of the Corporation's modified mandate, and found that the project still merited an award if suitable terms could be negotiated. If reader refers to the Business Plan described in the prior chapter, he will note that the true in situ shale technology was only a secondary priority for the Board. Given, however, that a letter of intent had been signed, that the project was ready to proceed, and that the amounts of assistance would be modest, the Board authorized the staff to proceed with further negotiations.

Potential problems were immediately apparent. It was clear that the amount of assistance would have to be significantly higher than that negotiated as part of the letter of intent. Peter Kiewit Sons, Inc., joined the project as a major sponsor in April, 1984, and undertook a redesign of the project, incorporating test results from the demonstration retort that had not been previously available. The redesign resulted in higher estimates for the project's capital costs. In addition, in the interval since the original terms had been negotiated, energy prices had declined, there had been an increase in transportation costs, and the need emerged for a contingency to deal with adverse tax law changes. These latter all led to the need for larger authority to cover price guarantee payments. Indeed, the total authority needed was to quadruple to $184.34 million by the conclusion of negotiations.

Would the Board swallow the increase? The prior Board had, of course, rejected a relatively modest increase in authority requested by the First Colony Farms peat gasification project one year after the terms had originally been agreed on the basis that a project "should not receive two bites of the apple." Moreover, the general expectation was that Congress's cutting of the Corporation's available authority would result in the downward negotiation of the amounts of assistance (as they did in all other cases) rather than upward.

In the event, the Board seemed to bite. One helpful feature of the larger amount required was the sizable contingency for unfavorable changes in the tax code (the major revision being considered by Congress at the time was likely to eliminate favorable treatment of items such as the energy tax credits and so forth). In this case, the contingency amounted to $36.51 million, a sizeable fraction of the total that might never be required. If it were not part of the package, the total authority amounted to $147.83 million, consisting of a $62.82 million loan guarantee and an initial $85.01 million price guarantee. (As in other awards, the amount available for price guarantees would increase as the loan was paid off). The price guarantee level was set at $55 per barrel escalated for inflation, a level comparable to the other shale projects under consideration. There was also a provision for profit-sharing for a period of fifteen years from the start of production.

Contract documentation proceeded apace. Final approval was scheduled for Board consideration on September 24, 1985. But out of left field came another key delay. On September 17, 1985, the U.S. Court of Appeals for the Tenth Circuit unexpectedly ruled on a case that had been pending for many years concerning the jurisdiction of Indian tribes over certain Utah State lands (*Ute Indian Tribe v. Utah*, 773F.2d 1087 (10th Cir. 1985)). This decision threw a cloud of uncertainty over the question of ownership of the lands and the possibility of the Indians claiming royalties, and so forth. The Board put off action until the picture clarified.

After considerable consultation with attorneys and the Utes, the Corporation concluded that any risks lay mostly with the sponsor and were probably acceptable. Again, the Board scheduled a meeting to consider final approval of the assistance. But it was not to be. The meeting was scheduled for December 17. On December 16, Congress voted to prohibit the Corporation from making further awards of financial assistance.

Of the variety of issues and problems that arose to plague projects being considered by the Corporation, Seep Ridge's experience was probably the most vexing.

Progress on the other Letter of Intent Projects

The above discussion described the substantial negotiating efforts expended with regard to four of the letters of intent. Two of the other projects that had negotiated term sheets were withdrawn from consideration after their sponsors were acquired by new entities that lacked interest in synthetic fuels. Kentucky Tar Sand was withdrawn when Texas Gas was acquired by CSX Corporation and HOP Kern Heavy Oil Project was withdrawn after Ladd Petroleum was acquired by GE. During this period, little progress was made with regard to the Northern Peat Project, which was obviously not a high priority for the Board.

Cathedral Bluffs was the last of the eight projects that had letters of intent pending when the reestablished Board became functional. The Board had explicitly indicated in the Business Plan that the modified in situ technology represented by the Cathedral Bluffs was a high priority. However, the Board indicated that it wished to see a surface retort other than the Union "B" included in the project in order to achieve the Corporation's diversity goals. In addition, the reduced emphasis on production and the reduced amount of obligational authority suggested that the $2.1 billion of authority negotiated as part of the letter of intent would have to be renegotiated sharply downwards.

The project's sponsors were willing to reconfigure the project in line with the changed circumstances and the Board's wishes. The first step was to determine how small the project could be and still prove out the technology and be economically viable over a reasonable commercial lifetime. The sponsors' staffs developed a series of options for the Corporation to consider. These varied the number of in situ retorts that would be operating at a given time, as well as different sized surface retorts. Discussions were focusing on options that would have required less than $1 billion of authority when the Corporation was forbidden by Congress from making any further awards.

THE PENDING SOLICITATIONS

Before the Board had lost its quorum in April 1984, it had issued three solicitations under which proposals were received in the interregnum between active Boards. The first two of these, the Coal-Water Fuel Solicitation and the Solicitation for Coal or Lignite Gasification or Liquefaction Retrofit Projects, have already been discussed in Chapter 8, which described all the targeted solicitations. The other solicitation under which proposals were received during the period when the Corporation was unable to act was the Fourth General Solicitation.

The Fourth General Solicitation was issued in February 1984 to allow all projects that had achieved a high degree of development a final opportunity to seek assistance. At that time, the Board was especially interested in coal technologies. Therefore, the solicitation had been structured to provide for two award categories, one for projects using coal from the Eastern Province or the Eastern Region of the Interior Province and proposing to produce at least 6,500 barrels of oil per day equivalent, and one for all other projects.

Seventeen projects applied to the Corporation under that solicitation[138]:

American Syn-Crude/Indiana Oil Shale, to be located in southeastern Indiana using the PETROSIX retort to produce 2,011 BOED.

Byrne Creek Power, to be located in Uintah County, Wyoming, using an in situ coal gasification process to produce 1,478 BOED of medium-Btu gas.

COGA-1, to be located near Springfield, Illinois, using the Texaco coal gasification process to produce 13,750 BOED of anhydrous ammonia.

Cottonwood Wash, to be located in Uintah County, Utah, using the Paraho retorting process to produce 8,547 BOED of shale oil.

Edwards Engineering, to be located in the Green River Formation area using the Edwards "Anaerobic" Metal Surface retort to produce 1,000 to 2,000 BOED of shale oil.

Enpex, to be located in Maverick County, Texas, using the Conoco FAST tar recovery process to produce 950 BOED of tar sand oil.

Forest Hill, to be located in Wood County, Texas, using an in situ oxygen fireflood process to produce 1,633 BOED of heavy oil.

International Hydrocarbons, to be located in Carbon County, Utah, using the Smith Extraction Process to produce 48,768 BOED of shale oil, electric power and ammonia.

Keystone, to be located in Somerset County, Pennsylvania, using the KRW pressurized, fluidized bed coal gasification technology to produce 6,620 BOED of fuel gas.

KILnGAS Cogeneration, to be located at East Alton, Illinois, using the KILnGAS coal gasification system to produce 1,742 BOED of low-Btu gas.

Louisiana Fuels Group, to be located at Lake Charles, Louisiana, using the British Gas/Lurgi slagging fixed-bed gasifier to produce 10,000 BOED of fuel gas to generate electricity.

Means Oil Shale, to be located in Montgomery, Menifee & Bath Counties, Kentucky, using the Dravo Traveling Grate Retort to produce 14,000 BOED of shale oil.

North Alabama, to be located at Murphy Hill, Alabama, using the Texaco Gasification Process to produce 7,420 BOED of methanol.

PR Spring Tar Sand, to be located in Uintah County, Utah using the Solv-ex process to produce 4,180 BOED of oil from tar sands.

Scrubgrass, to be located in Scrubgrass Township, Pennsylvania, using the Koppers-Totzek coal gasification process to produce 3,917 BOED of gasoline, propane, and butane.

Sweeney, to be located at Sweeney, Texas, using the KRW pressurized, fluidized bed coal gasification process to produce 4,000 BOED of fuel gas.

Utah Methanol, to be located in Wood County, Utah, to produce 330 BOED of methanol from coal using a then-unspecified process.

There were many familiar faces in this list. One, Forest Hill, applied even though it already had a letter of intent from an earlier solicitation so as to give itself another option should that letter fall through for any reason. The majority of the others had applied in earlier solicitations and had continued their development as encouraged by the Corporation, although a handful were new.

The Solicitation was structured with relatively strict eligibility criteria. The projects had to demonstrate an advanced degree of engineering development with a Level II/III design, and had to show a financial commitment from the sponsors of at least 60 percent of that which would be needed to build the project with assistance from the Corporation. During the interregnum between Boards, the staff review had shown that only six of the seventeen projects could be considered qualified against those strict criteria. The other projects could not, however, be dropped from consideration until the quorum of the Board had been restored.

Then, the board, as recounted above, developed some new programmatic criteria, as reflected in the Business Plan. The qualified projects had also to be evaluated against those criteria. During early 1985 the Board acted to drop the projects found not to be qualified or found not to meet the Board's programmatic criteria. At that point, only four projects remained. Forest Hill, one of these, chose to remain in the Third Solicitation. The three remaining qualified projects were: American Syncrude/Indiana Shale Oil, Keystone, and Utah Methanol.[139]

Over the ensuing months, the staff worked closely with all three projects to ensure that the projects were truly ready to proceed to a completed assistance agreement and to begin to scope out the amount and form of assistance that would be required for a viable project. A capsule view of the efforts on the three projects follows:

The American Syncrude/Indiana Oil Shale Project. This project, intended to extract about 2,000 BOED of oil from eastern Devonian shale, was sponsored by a group including American Syn-crude Corporation, Stone and Webster Energy Corporation, Petroleo Brasileiro, S.A., Cives Corporation, Sulzer Brothers, Inc., Petro-Chem Development Co., Inc., Perry Gas Processors, Inc., and McNalley Pittsburgh, Inc. Many of the projects applying to the Corporation had experienced major difficulty in raising the necessary capital. The project had the added complication of having one of its major sponsors—Petroleo Brasileiro—an entity from a nation that had complicated foreign exchange restrictions. Determining that the project could meet the tests established by the Solicitation was not an easy task. In addition, the Board was ambivalent about the project; eastern shale was a secondary priority. On the other hand, the project would have added to the diversity of oil shale technology, and the technology would have been applicable to western shale as well.

The Keystone Project. This project had two substantial attractions for the Board. First, it was sponsored by the Signal Companies, Inc., which had the financial wherewithal to undertake the project, thus avoiding the ubiquitous equity problem (although they could decide to bring other partners into the project). Second, the KRW fluidized bed gasification process was a high priority. All the other coal projects supported by the Corporation or DOE had employed either fixed-bed or entrained-flow gasification. The focus of negotiations was reducing the scale and amount of assistance from that initially envisioned by the sponsor to that corresponding to the new funding realities facing the SFC. The original proposal was for a project producing approximately 6,500 BOED of medium-Btu fuel gas to be used in a combined-cycle power plant to generate electricity. Negotiations had pretty much closed in on a facility of the lesser size of 3,500 BOED as meeting the needs of the sponsor as well as the Corporation.

The Utah Methanol Project. This project was sponsored by the Questar Synfuels Corporation, a wholly owned subsidiary of Questar Corporation. The proposal had initially left the gasification technology as unspecified, but quickly identified a Mountain Fuels entrained flow (dry feed) coal gasification process. The dry feed aspect was of particular interest to the Corporation since it held the promise of higher efficiencies than the slurry-feed entrained flow processes being used in the Texaco and Dow projects. The principal issue concerning the project was whether a 330 BOED project could be considered to be commercial scale (and, thus, meet the requirement of the Act for eligibility). The staff was evaluating the argument that the technology scale being proposed would deal with virtually all of the new technology risk, such that future replication of the technology could proceed as easily as would be the case for the other projects being assisted by the Corporation, and that the project would have sufficiently viable economics to allow a "normal" commercial life.

NEW SOLICITATIONS

After the reconstituted Board developed its Business Plan determining the kinds of technologies and projects it felt were essential for accomplishing the redefined mission of the Corporation, and they had reviewed the projects still pending action, they were able to identify the targeted areas for which no candidate projects were in sight.

Consequently, the Board chose to issue two last solicitations to attract the remaining projects. Moreover, this was not a perfunctory effort given that they had a reasonable expectation that viable projects were likely to apply. These solicitations, which were discussed at length in Chapter 8, were: 1) Solicitation for Eastern Province or Eastern Region of the Interior Province Bituminous Coal Gasification Projects, and 2) Solicitation for Projects to Produce Synthetic Fuels by Mining and Surface Processing Tar Sands.

They were issued in May and June 1985, respectively. They did attract likely prospects for assistance. In the coal solicitation, the Keystone (also applying under the Fourth General Solicitation), COGA-1, and Virginia Power projects represented projects with substantial potential for being awarded assistance. Under the Tar Sands Solicitation, the PR Spring and Sunnyside projects appeared to meet the criteria for successful projects.

The reconstituted Board had quite a year! They had:

Concretely recast the Corporation's mission to implement the legislation of 1984 by formulating the Business Plan and preparing the Comprehensive Strategy Report for Congress.

Completed assistance agreements with Parachute Creek and Forest Hill, and came within a hair's breadth of accomplishing the same with Great Plains and Seep Ridge.

Acted on the pending projects under the Fourth General Solicitation to focus on American Syncrude, Keystone, and Utah Methanol.

Reshaped the Cathedral Bluffs project with the sponsors and substantially advanced negotiations with this last letter of intent project.

Closed out the Coal-Water Fuels and Retrofit solicitations.

Issued the targeted coal and tar sands solicitations and acted on the submitted proposals to begin working with the COGA-1, Virginia Power, PR Spring, and Sunnyside projects.

In short, the Board had brought the program to the verge of sensible success. The Corporation had either completed agreements, letters of intent, or excellent prospects under review for projects that represented virtually a full array of technologies and resources that met the diversity goals of the Act. Had these projects all gone ahead, the country would have been assured of having the most cost-effective technologies for producing synthetic fuels available, should they be needed in the future. Moreover, the program could be achieved within the reduced funding stipulated by Congress the previous year. The Comprehensive Strategy Report estimated that the entire program would cost in the range of $6.5 billion to $9 billion expended over about a thirteen-year period. In constant 1985 dollars the amount would have only been about $3 billion. This amount of real national resources would have been less than had been spent by the federal government over two or three decades to work on synthetic fuels R&D, but which had provided no production capability (immediately available or potentially replicable). And certainly, it was a far cry from the $20 billion initially appropriated under the Energy Security Reserve.

They also, as we shall see in the next chapter, fought for the Corporation's life in Congress, but failed. So, ironically, just as the Corporation was just ready to complete the mission that Congress had assigned it with such fanfare the prior year, that same Congress turned tail and terminated the Corporation on December 18, 1985.

Let's see how that unanticipated turn of events unfolded.

XII

Eroding Political Support

The previous chapter described the Corporation's accomplishments that brought it close to accomplishing its revised mission of providing the nation with a diverse array of technological alternatives for exploiting its vast solid energy resources to provide alternative fuels. Yet Congress was about to turn its back abruptly on what it had strongly promoted for the previous five years. Why? And how could this come to pass when the leadership of both parties in both the House and the Senate staunchly supported the Corporation and its modified mission? The fundamental cause was that public attention had turned elsewhere and political supported waned as energy circumstances had improved.

This alone, however, could not have explained the events that led so suddenly to the premature termination of the SFC. Indeed congressional termination of any program is a rare event. In the case of the SFC, however, there had been significant opposition to the Corporation's mission from its inception and that opposition gained strength with the support of sympathetic Congressional committees as public support ebbed. The orchestrated efforts to undo the Corporation viewed as a whole can be used as a case study of how motivated ideological movements can draw on congressional allies and sympathetic members of the media to achieve unexpected ends.

To convey how this unusual set of events unfolded, this chapter will describe: (1) the ongoing opposition to the SFC, (2) the criticism leveled at the Corporation, (3) events in Congress, and (4) media coverage. In addition, this treatment provides background on SFC administration, relations with Congress, and the media not covered elsewhere in this history.

THE OPPOSITION

Chapter 3, which described the crafting and passage of the Energy Security Act, indicated how the support of philosophically disparate interest groups was achieved by providing monies for renewable energy sources and conservation on one hand and a synthetic fuels program on the other. But the supporters of the former accepted their benefits from the compromise without being reconciled to the purposes of the latter. Moreover, there were groups unhappy with both elements of the compromise.

Conservatives

A number of conservative congressmen, opposed to such a large governmental intrusion into energy markets, figured prominently among the latter group. These attempted periodically over the years to curtail the Corporation's mission, reduce its funding, or eliminate it entirely. For example: on February 2, 1982, Representative Hank Brown (R-CO) introduced H.R. 5404, a bill to abolish the Corporation; Representative Tom Corcoran (R-IL)[140] on March 29, 1982, introduced H.R. 5977, a bill which among other things would have terminated the Corporation by September 30, 1984; and in April of that year, Senator Armstrong (R-CO) introduced S. 2362, a bill to abolish the Corporation. But with the leadership of both houses strongly supportive of the SFC, these bills never came to a vote.

Other congressmen, notably John Dingell (D-MI), strongly opposed the idea of creating a quasi-federal corporation given very large funding autonomy that was outside of Congress' annual appropriations process. Congressman Dingell was a formidable antagonist given that he chaired the powerful Committee on Government Operations/ Subcommittee on Oversight and Investigations (O&I).

Other significant opposition centered in environmental groups who brought ideological fervor to the opposition to the entire synthetic fuels program, not just the SFC.

Environmentalists

Environmental groups had spearheaded attacks on the SFC even before the Corporation was operational. Their mindset had little to do
with how well the SFC was managed but instead was based on ideological opposition to rapidly growing economies that consumed
ever-larger amounts of natural resources. As such, these groups
went beyond middle-of-the-road values that favored a cleaner environment. Rather, their agenda was one consistent with the writ-
ings of E.F. Schumacher in *Small is Beautiful,* a call for smaller institutions, a move back to the land, adoption of simpler technologies,
and to an environment "more people-oriented."[141] Only conservation of energy, and solar and biomass resources were acceptable;
nuclear energy and large synfuels plants were anathema. This sort
of battle was, of course, not just about synthetic fuels. The same sort
of issues arose in other areas, such as drilling for oil in Alaska and
off-shore, building dams, the siting of conventional power plants,
using nuclear energy, acid rain and so forth. These views persist a
quarter of a century later, with the focus now on "global warming."
    This outlook is based on the conviction that the world is rapidly
depleting vital natural resources and that without radical change
leading to *sustainable growth*, calamity ultimately awaits us. An illustrative view of these beliefs can be seen in a work entitled *Seven
Tomorrows,* which drew on work at the Stanford Research Institute
undertaken by a Futures Research Group. [142] The authors suggested
that without radical reordering of society, the United States would
face circumstances similar to what they termed the "Official Future:" urban homeless increasing tenfold in the eighties, a considerable rise in lung-related diseases, polarization of labor unions,
featherbedding in declining industries, a rise in suicides among the
middle-aged, an oppressive legal system, and so forth. They implied that the *best* economic growth scenario envisioned by their
studies could not endure. They suggested that its inevitable difficulties stem from a "failure to appreciate complex and interactive
consequences [which] is partly a function of people's faith in offi-
cialdom: despite visible deterioration of parts of the environment,
despite exponential growth rates that threaten imminent shortages
of resources, an abiding confidence in technical expertise [that] allows most individuals to proceed with business as usual".

With such an apocalyptic view, factual details about this synthetic fuel plant or that, or the program as a whole were irrelevant; they simply had to be opposed—along with many other features of a modern industrial economy. On July 13, 1990, an editorial "Environmental Balance" in *The Wall Street Journal,* commented on a quote of the first President Bush: "But on the environmental extreme, they don't want this country to grow. They don't want to look down the road at the human consequence of men and women thrown out of work and families put into a whole new state of anxiety. And I as President have to be concerned about that, as well as being a good custodian, a good steward for the environment. We cannot govern by listening to the loudest voice on the extreme of the environmental movement." The editorial went on to say "We suspect that Mr. Bush is not the only person who is beginning to feel that his good-faith efforts on environmental matters will never satisfy movement activists whose agenda is increasingly taking on the trappings of religious fanaticism."

While many mainline environmental groups represented members embracing conventional goals of clean water and air, wetlands protection, and protection of endangered species, their behavior vis-à-vis the Corporation reflected more extreme environmental outlooks. Who were these groups? While antagonism to synthetic fuels was present across the board, the Sierra Club, Friends of the Earth, and The Environmental Defense Council were prominent. There were also numerous state and local groups that opposed individual projects, such as the Kentucky Audubon Council, the National Wildlife Federation Carolina Wetlands project, the Powder River Basin Resource Council, and Save Our Cumberland Mountains, among others.

The active opposition of environmental constituencies coalesced around the Environmental Policy Institute, located in Washington, D.C. And, everywhere the SFC went, one of the representatives of the Institute went as well. They appeared at virtually all of the SFC Board meetings, which were open to the public. They staged events to embarrass the Corporation and call attention to their cause such as handing out antagonistic statements along with the Board's statements at SFC press briefings, and on one occasion, wandering through the audience handing out chocolate candy in the shape of gold coins to suggest that the SFC actions were simply throwing money away.

They were equally ubiquitous whenever a briefing was organized for congressional staff, such as by the Congressional Research Service. At these meetings the Institute's representatives would be present to attack the SFC and disrupt the discussion. What they lacked in scientific data and intellectual rigor, they made up for with perseverance and presence.

During the early years of the Corporation, the individuals most in appearance were Robert L. Roach and Rick Young. Roach was director of the Institute's Synthetic Fuels Assessment Project, and Young was a research associate at the Institute who later became a staffer on Congressman Synar's Subcommittee. He was the liaison of the subcommittee with the Corporation with regard to all of its hearings aimed at embarrassing the Corporation and its voluminous requests for information. Roach also later became a staffer, in his case for Howard Metzenbaum, a Democratic senator from Ohio and the SFC's main opponent on the Senate Energy Committee. There were close links between the antagonistic environmental groups and those congressmen most active in attacking the SFC program.

Given that at the time no plants had been built or operated, their attacks had no grounding in any specific environmental impact per se; rather they focused on highlighting unavoidable costs and uncertainties inherent in pioneer projects. The best illustration of these can be seen in two publications put out by the Environmental Policy Institute in 1982: "Dreams into Dollars: A Review of Five Candidates for Synthetic Fuels Corporation Assistance" and "Distant Dreams: The Synthetic Fuels Corporation's Search for a Commercial Industry." These attacked the first projects that passed the SFC's maturity review. "Dreams into Dollars," the first, concerned five seeming contenders: North Alabama, Paraho-Ute, Hampshire, Wycoal, and Tennessee Synfuels. While most of these were serious projects that resubmitted proposals for Board consideration, only Hampshire passed the Strength Reviews (discussed in earlier chapters) making it eligible to enter negotiations for financial assistance. When it became apparent that these targets were not leading contenders after all, the Institute issued "Distant Dreams," which in turn attacked a new set of applicants: Memphis, Calsyn, First Colony, Breckinridge, and Hampshire (again).

Virtually none of the criticism dealt with environmental issues, other than noting that the permitting processes for some of the projects were still uncompleted. They focused instead on the uncertainties associated with individual projects, e.g., incomplete equity sponsors and efforts to reduce the size of some of the projects from what had been originally proposed.

The core of the attack, however, was against the existence of the SFC and its funding:

> As federal fiscal policies have been scrutinized as never before, we have been made painfully aware of the fact that government expenditures, whether direct, off-budget or deferred, impose costs ...
>
> The SFC projects that it will outlay $66.17 million in administrative costs in FY 81 - FY 83. The Congressional Budget Office (CBO) estimates that up to $186 million in administrative costs alone may be saved through the years FY 83 and FY 87 by *terminating the SFC* (author's emphasis).
>
> Price guarantees made by the SFC will inevitably result in massive outlays by the federal government in the late 1980's and 1990's, and perhaps even earlier, when the products of synfuels projects are not cost competitive.

But Congress understood that the program would be costly when it authorized $88 billion to achieve a national security objective. Given their goal of protecting the environment, these groups should have supported the development of a detailed scientific record of the environmental impact of such technologies. Earlier, when the Energy Security Act was being drafted, they had highlighted speculative environmental uncertainties associated with the output of the new first-of-a-kind facilities, for which there was no track record. As a result, Congress required all of the plants funded by the SFC to carry out extensive monitoring of all effluents—both those regulated by permits and those unregulated. If environmental groups had any reason to believe that some substances, regulated or not, might be hazardous, then project sponsors would monitor effluents to determine amounts and impact. Then, if such substances were detected, future regulatory agencies could require control technologies be employed if and when synthetic fuels plants were to be replicated on a wider scale. Support from environmental groups for any aspect of the program, however, was not forthcoming.

PRINCIPAL CRITICISMS OF THE SFC

The most consequential and damaging criticisms of the SFC arose within Congress, drawing on missteps by the Corporation, on program delays outlined in earlier chapters, and on exaggerations of corporation operations that were exempted from some federal agency regulations. These elements were woven into a narrative of a "scandal-ridden corporation" that was extravagant in its spending and that funded unworthy projects. The following sections analyze the elements of this narrative, laying out the charges and providing a context to judge their merit.

The Corporation's Leadership

During the SFC's brief lifetime, it had only two presidents—Victor Schroeder and Victor Thompson—and these only during its first years of operation. Let's review each of their tenures in turn and examine the circumstances under which they each left that office.

Congressional criticism of Victor Schroeder largely drew on his lack of background in the energy industry, on some issues that were initially raised by internal SFC reviews, and from internal disagreements among SFC Board members regarding solicitation strategy.

It was widely recognized that Schroeder was made President of the Corporation because he held the full confidence of the Chairman Ed Noble. They had worked together for many years in undertaking large commercial real estate development entailing complex project management, requiring judgments of the financial capability of participants and the financial viability of projects. Noble himself came from a family with sizeable energy interests in natural gas and was convinced that, supported by a capable SFC staff incorporating expertise in synthetic fuels technologies, Schroeder was the man to manage the Corporation in accordance with Noble's (and the Board's) policy direction. The close working relationship between the two men that predated their time at the SFC, however, was characterized by some Congressmen as cronyism despite no evidence that their financial affairs were linked at this time.

Unfortunately, early on Schroeder made a couple of missteps that, while perhaps commonplace in the private sector, were found questionable in a publicly funded entity. The first issue had to do with expenses Schroeder had claimed when relocating from Atlanta, Georgia to assume the full time position of President of the Corporation. He claimed reimbursement for $19,500 for the real estate broker's commission on the purchase of his new house, and for $4,770 in mortgage rate differential payments. As a general matter, these types of payments were considered appropriate under the Corporation's relocation policy, but the way in which Schroeder used them was not.

With regard to the commission, a report prepared by the Inspector General stated "… such commissions are normally paid by the seller; and, neither the old draft policy [of the SFC on relocation expenses] nor the new adopted policy provide for real estate commission reimbursement on the purchase of a house. The $19,500 on the surface appears to be additional consideration toward the purchase cost of the house. The employee did not sell or pay any selling commission on his former residence. Therefore, no commission seems to be reimbursable under the relocation policy." A further review by the Office of the General Counsel reached a similar conclusion.

The Inspector General's audit also questioned three annual payments of $1,590 claimed by Schroeder as mortgage interest rate differential payments. The Corporation's relocation policy in effect provided for payment for interest rate differences between old and new rate times old mortgage dollar balance for a three-year period with payment to be made on an annual basis. The Inspector General concluded that the three claims were open to question because Schroeder did not sell his former residence or forego the benefits of a lower interest rate. Schroeder subsequently reimbursed the Corporation for the above amounts before the issue received any external attention.

A second area of criticism related to making non-competitive procurement awards in the SFC's first year and three quarters of operations during which it awarded sole source contracts totaling $623,465. Two contracts in particular received congressional scrutiny: i.e., with two individuals whom Schroeder had engaged in past years during his work on major real estate development projects and in whom he had special confidence (but no personal business ties). One individual performed the work for space layout and interior design for the four floors occupied by the Corporation in a new building (involving about $100,000 in fees). The other individual performed work on the "communications problem" that was provoking the Board members, involving about $21,000. One Subcommittee observed that the former involved work that other outside firms could also have done, and the second could have been performed by existing SFC staff.

This being said, the SFC's procurement record on this matter matched or was superior to that of other federal agencies. As a practical matter, if a small award is to be made to an individual for a modest task, it is impractical and not economic to undertake a competitive process. Establishing a competitive procedure, soliciting proposals, and reviewing results would consume more resources than could conceivably be saved for a modest effort entailing less cost than that of a full-time position. In that light, the Corporation actually had a superior performance vis-à-vis federal agencies in awarding contracts competitively. Indeed, a Subcommittee Report to be discussed later stated "Overall, more than 80 percent of the Corporation's contract dollars were awarded competitively in fiscal 1982—a commendable record." So, this issue would seem to be of small interest value, except for the attention it received from members of Congress and the media afterwards.

The most damaging criticism of Schroeder resulted from disagreements he had with other board members when they became the subject of a congressional hearing. While there were a number of contributing elements, the disagreements seemed to grow out tension between Schroeder and Robert Monks, one of the outside directors. Monks had been devoting considerable time to the Corporation's affairs and believed that its program was not moving quickly enough and was not approaching likely sponsors in the most effective way. Schroeder, on the other hand, believed that the solicitation process the Corporation was following was slowly but surely bearing fruit. In any event, the tension between the two views was to result in a management explosion damaging to Schroeder and to the Corporation.

In a nutshell, Monks believed that only large corporations with "deep pockets" would be successful in building and operating large pioneer plants without undue expenditures by the government. He believed that the solicitation process being followed was too bureaucratic and was not an effective way to engage with the most appropriate of the likely sponsors. He noted that the Act did permit the Corporation to enter direct negotiations with sponsors, if the solicitations failed to produce results on a competitive basis. Moreover, at this time, it was clear that the first solicitation would result in no awards at all.

At Monk's urging, the Corporation issued the first of the targeted, competitive solicitations that would have allowed direct negotiations. (See the discussion in Chapter 8.) While that approach did lead to direct negotiations on a project sponsored by Arkansas Power, it did not result in an award. Schroeder felt that making awards outside a competitive solicitation was politically dangerous, especially when the solicitation process was showing increasing signs of producing results. Surely, one would think, these views were amenable to a working compromise. But in the final analysis, Monks seemed not to respect Schroeder's capabilities and wanted him to resign.

Monks brought the conflict to a head with two actions. First, he handed the Chairman an undated and unsigned letter (indicating concurrence from Howard Wilkins and Jack Carter, two of the other outside directors) prior to the May 26, 1983, meeting of the Board. In this letter, they called for the chairman to "make suitable arrangement for termination of the service of the incumbent President of the SFC."

Second, Monks, as a wealthy supporter of the Republican Party from Maine, drew on his relationship with Senator William S. Cohen (R., Maine), Chairman of the Senate Committee on Governmental Affairs, to have that Committee investigate the management of the Corporation in a formal hearing.[143]In his introductory remarks at the hearing, Senator Cohen noted that he had sponsored Monks for a seat on the SFC Board. Further, the Bangor, Maine, *Daily News* published an article just before the hearing with the title "Monks, Cohen Allied in Synfuels Power Struggle."

The central focus of the hearing was a matter described as "vote trading", which allegedly occurred on March 23, 1983, when Schroeder met with Mike Masson to review items on the agenda for an upcoming Board meeting. This meeting was partially the result of outside directors desiring closer communications with the full time directors. Indeed, one of the items on the agenda of the Board meeting and discussed at length between the two men was Schroeder's proposal to deal with the perceived communications shortfalls. The meeting between Schroeder and Masson lasted several hours.

One item, however, not related to SFC business arose at this meeting, which took place late in the evening on their own time. Masson's description of the interchange, which was provided in testimony, was that the discussion grew out of Masson's role as the head of an engineering firm in Phoenix. Masson stated [in congressional testimony]:

> I did indeed ask Vic Schroeder ... if he in fact knew anyone at Mobil Land Corp. because I had heard that they bought approximately 2,000 acres of raw land in the desert near Phoenix, Arizona. I would like to confirm that. He said earlier [at a prior meeting] as a matter of fact, I have a personal friend of long standing who is in charge of Mobil Land Development. I said, great, would you call him or put me in touch with him, because I would like to know. The night of our meeting, I very clearly reminded Vic. I said, 'Vic, by the way, would you please follow through and call the guy and see if they have land in Arizona?' Subsequent to that, at some time, 2 or 3 weeks later, he wrote me a memo, indicating that here's a man's name at Mobil Land; he is expecting your call; you can call him and ask him. Since that time, I have not had the time or the interest to follow through.

Later he testified, "I am sure, if I hadn't been so lazy after I found out Vic Schroeder had a personal friend there, I could have looked in the phone book and found someone at Mobil, and found out that way."

How could this innocuous interchange be the subject of a "scandal" and a congressional hearing? It transpired that a third party was present at the discussion, but not a participant: Don Thibeau, Executive Assistant to the Chairman. One of his chief roles involved liaison with the outside Board members, and he frequently sat in meetings of this kind. After the several-hour meeting between Schroeder and Masson, which covered an array of topics, he felt that an arrangement had been reached. Masson would support Schroeder's proposals, and Schroeder would provide the name of someone at Mobil Land. Others characterized this as "vote trading." Being uncomfortable, Thibeau reported the conversation to his immediate boss, Noble, but he also mentioned it to Monks and Wilkins.

The Subcommittee devoted much of its two days of hearings taking testimony from Schroeder, Masson, Thibeau, and Noble regarding this issue. The Subcommittee Report found

> While the Department of Justice has concluded that insufficient evidence exists to warrant further investigation or prosecution for criminal violations, the Subcommittee believes that this conversation, nevertheless, raises serious ethical concerns. Of prime concern to the Subcommittee is the two Directors' apparent lack of sensitivity to the appearance of impropriety. ... As public officials, these persons must scrupulously avoid not only actual, but also the appearance of, impropriety.

The Report went on to cite the Corporation's ethics policies that forbade the acceptance of gifts, gratuities, entertainment and favors from those who have or seek business with the Corporation. Since neither man asked for nor received anything from Mobil, and even if they had, Mobil never had any business dealings with the Corporation, their exchange would seem to be irrelevant to Corporation policy. Nonetheless, the Report went on to say: "The Subcommittee finds that the agreement to have Mr. Schroeder contact Mobil Land Development Corporation on behalf of Mr. Masson may violate the spirit, if not the letter, of this SFC policy." (Schroeder, of course, did not contact Mobil, but simply gave Masson a name.)

Continuing, the Report stated, "The Subcommittee recommends that the Board of Directors establish disciplinary procedures for violations of the Corporation's ethical standards." In addition, without adding to information previously raised by the SFC Inspector General, the Hearing Report observed that "the $19,500 reimbursement [for moving expenses discussed earlier] was improper and should not have been made ... [and] the Subcommittee endorses the Inspector General's opinion that adequate grounds exist to question the propriety of the claims for $4,770 ..."

As a consequence, Schroeder concluded that in the best interests of restoring harmony on the Board, he would step down as president but remain as an outside Board member. Indeed, that seemed to be the view of the Subcommittee. The Report stated "The Subcommittee believes that Mr. Schroeder's decision to resign as President reflects his concern for the best interests of the Corporation. His resignation will allow the Directors to choose a new President who will enjoy their full support and confidence. That should mend the division in the Board and permit the Corporation to focus its energies on the synfuels investment program."

The political process in Washington, however, does not just move on. Members of Congress and the media made this the beginning of the "scandal-ridden" Corporation. Numerous references were made to shady vote-trading and to Schroeder somehow slipping contracts to business associates, implying a gain to himself (although he had no business relationship to the consultants in question). After all, a congressional hearing was held and he was forced to resign. Therefore, he must have been guilty of something significant. He would show up on lists of those forced from office in *The Washington Post* whenever they wished to make the most of the "sleaze" factor under the Reagan administration.

Finally, having achieved his objective of forcing Schroeder from the corporation, Monks resigned from the Board a few months later to go on to other interests.

After Victor Schroeder was pressured into resigning in the late summer of 1983, the Corporation was without a president. At the same time Jimmy Bowden resigned as executive vice president. Chairman Noble had made it clear from the outset that his focus was the overall mission and policy, not the day-to-day management of the Corporation. While he adopted an interim measure of creating two supra-vice presidents, he continued to examine the Corporation's options for filling the president's position. In early 1984, he offered the position to Victor Thompson.

There were a number of obvious advantages to this move. Thompson had had extensive financial experience as the Chairman of the Utica Bank Shares Corporation, a bank with approximately half a billion dollars in assets that was located in Tulsa, Oklahoma, and with experience in the energy industry. He was already a member of the Board of Directors of the SFC, and importantly, he had been one of Noble's bankers for many years, thereby having gained the confidence of the Chairman.

Accepting the position was not an easy move for Thompson. It required him to relinquish a three-year contract with Utica and an opportunity for full vesting in its retirement plan. Moreover, he had to sense that the enemies of the Corporation were on the move and would be attacking him as well as the Corporation during his tenure. It is a tribute to Victor Thompson that he accepted the position anyway.

While the attack on Victor Schroeder was initially a tactical one by individuals largely friendly to the Corporation, the attack on Victor Thompson was orchestrated by the Corporation's chief antagonist in the House of Representatives, John Dingell. Thompson assumed the presidency on February 29, 1984. By April 3, 1984, he was already the subject of a congressional hearing by Dingell's Oversight and Investigations Subcommittee.[144]

Because just four weeks of tenure as president gave little basis for an oversight hearing, Dingell chose to attack him on a past event that had occurred at Utica Bankshares. The starting point for this Committee's examination of Thompson was a Securities and Exchange Commission Report on Utica Bank regarding reports that had been filed a year and a half previously, tangentially associated with the prominent Penn Square Bank failure that occurred almost two years prior to the hearing.

For context, the reader should be aware of the events in the banking community beginning in 1982 (the period under review). With the deep recession of 1982, and falling oil prices, many banks in energy producing states (such as Oklahoma) experienced severe financial pressure from defaulted loans, with the consequence that many banks failed. Indeed, by 1989, there was only one major bank in Texas that had not been taken over by outside interests in reorganizations.

The Utica Bank was not immune to this environment. It had bought some of the loans syndicated by Penn Square as had many banks (at an eventual cost to Utica of $7.7 million). More importantly, it began to experience growing numbers of non-performing loans during 1982, a matter subject to SEC regulations and general accounting standards, i.e., determining when a bank should draw on reserves to cover non-performing loans. An SEC Report reviewed the bank's 10-Q Forms for the second and third quarters of 1982, concluding that the bank did not provide for losses promptly enough and should have taken action one quarter sooner than they did. The Report cited the bank's Chief Financial Officer who had the responsibility for these decisions, but did not mention Victor Thompson. Thompson, as chairman, however, was also obliged to co-sign them. He argued that he relied on the staff of the bank to carry out the required procedures in the preparation of material and did not routinely personally investigate whether the procedures had been properly executed.

The questioning proceeded to make an issue of whether Thompson had kept the Chairman of the SFC aware of the specific events at the Utica Bank—particularly the fact that the SEC was evaluating the 10-Q reports. Thompson stated that he did not because he was informed that the SEC Inquiry was confidential and he was not a subject of the inquiry.

Nonetheless, in an attempt to create the appearance of an ethics issue, Dingell called Owen Malone, the SFC Ethics Officer, to the stand. But Malone countered, "Mr. Chairman, there's no specific requirement in our policy concerning the reporting of lawsuits, the reporting of allegations and so forth. ... They are to avoid situations or actions that involve conflict of interest type conceptions."

Afterwards, Malone performed a follow-on review of Thompson's obligation to inform the SFC Board of the SEC Inquiry before the Board elected him to the Presidency. Malone concluded that while Thompson was obliged to keep the specific facts of the SEC Inquiry confidential, there was no legal constraint on him to preclude informing the Board of the existence of an inquiry. The accompanying memorandum from the Corporation's Legal Services Group concluded "Our review of general principals of corporation law governing the duty of a director to his corporation, and the analogous principles of the duty of a trustee to his beneficiary, suggests that Mr. Thompson had, under those general principles, a duty to raise the pendency of the SEC investigation to the Board's attention prior to his election."[145]

In a press release, Chairman Noble stated "After considering the materials and opinions presented to us, Vic Schroeder, Jack Carter and I believe that Mr. Thompson erred in not disclosing the ongoing SEC investigation prior to his election as Corporation President on February 16. We do not believe, however, that this is the kind of act that causes us to lose confidence in Vic Thompson's ability to contribute to the Corporation's effectiveness."

In the hearing, other lines of questioning were pursued. One important point was whether Thompson had personally benefited from the delay in the bank's making provision for the loan losses in reporting losses that led to a decline in the value of shares in Utica Bank. Thompson, however, testified that during this period he was buying rather than selling shares in the bank.

So, having unearthed no evidence in hours of hearing other than the fact that Thompson had not informed the Board of the pending SEC inquiry, Dingell had to draw on *appearances* the Chairman had created. He concluded this section of the hearing with the following statement:

> But, it is my earnest counsel to you that you should retire from your present position, both as a member of the board and as the chief officer of the corporation, at your earliest convenience, because it is my view that this Committee is going to maintain a continuing and active interest into the defects that we have observed in your performance of your office, your failure to adequately disclose your prior activities. And I must observe with considerable distress that

your performance this morning, your answers to the questions, your impeccably bad memory, bring me irretrievably to the firm conclusion that you are either incompetent to serve in this capacity or you are ethically unsuited to perform the responsibilities.

Victor Thompson did not wilt under this attack, and seemed to have no intention of resigning. Unfortunately, there was one more "appearance of conflict of interest" about to break on the scene, which was too much for him to withstand.

A conflict of interest issue arose the same month as the hearing just discussed. Chairman Noble had been called to the office of Congressman Loeffler (from Texas) who was inquiring into the treatment of one of his constituents, the Chapparosa Project, in the Corporation's solicitation process.[146]

Having discussed the circumstances of the Chapparosa Project, the Congressman alluded to the competition between that project and another similar project competing in the same solicitation, the ENPEX Project. He seemed to be concerned about the types of pressure that ENPEX could bring to bear on the situation. In the discussion, he mentioned that the Belton Kleberg Johnson interests of Texas had a small share of the Chapparosa Project and also owned a share of the office building in which the Utica Bank was located. He mentioned further that Vic Thompson had asked the owners of the building whether they would be interested in investing capital in the bank (see the above discussion regarding to the circumstances of banks in the Southwest).

As soon as the meeting ended, the Chairman discussed the matter with other Board members and with the General Counsel. The concern was that if an investor in a project in front of the Corporation were to make an investment benefiting Thompson, he would have to recuse himself from voting on matters associated with that project. The situation that emerged was that Thompson had, in a general effort to increase the bank's capital, inquired of the owners of the Utica Bank's office building whether they would be interested in investing equity in the bank. They replied that they were not.

That was the end of the matter as far as Thompson was concerned. However, upon further inquiry by SFC staff, he allowed as how he was generally aware that the interests owning the building had a small share in the Chapparosa Project, which had applied for assistance from the Corporation. However, as determined by a review triggered by the Chairman after the Loeffler meeting, Thompson was not precluded from engaging in business with entities seeking assistance from the Corporation, but he would have to avoid participating in decisions that could benefit such entities. In this case, he had voted on a matter involving the Chapparosa Project in the autumn of 1983 after he had invited investment in the Utica Bank.

Being questioned on the matter, he responded that a recusal made no sense. The parties had indicated that they had no desire to invest, and the matter was closed with no basis for a conflict of interest to exist. Moreover, the matter on which he voted was to determine that the project had *failed* to meet the requirements of the applicable solicitation and should be removed from consideration and would not be eligible to receive assistance.

The Ethics Officer review concluded that had Thompson reported his discussions with the Belton Kleberg Johnson Interests to the ethics officer, there would have been no possibility of a conflict. As it was, he was guilty of action that could have an appearance of a conflict so that he was technically in violation of the Corporation's ethics policy that precluded actions that could be interpreted as having a conflict of interest.

This conclusion, coming on top of the Dingell inquisition a few weeks earlier, was too much for Thompson. He tendered his resignation to the Board:[147]

> Gentlemen:
> This is my resignation as President of the United States Synthetic Fuels Corporation effective immediately. I have some comments to make.
> After experiencing the event chaired by Congressman Dingell, it is abundantly clear why people are reluctant to give more of their time, knowledge, and experience to foster a more efficient and stronger government. Chairman Dingell asked for my resignation, which I withheld at the time perhaps because I was stunned. It is unnecessary to repeat his comments regarding me; however, I was sure you and

the corporation had known me long enough to realize two things—one, that I was a human being quite capable or errors in judgment and the other that I was trustworthy and honest, possessing a strong sense of commitment and integrity.

I want to apologize to the Board and the staff for my believing the word 'confidential' means confidential. Incidentally, my dictionary says 'trust' is a synonym for confidence. It also says confidence is a 'relationship of trustful intimacy'. It is ironic when one agency uses the word 'confidential' and by accepting it at face value, it violates the ethics of another government entity. What the author of the letter from the SEC to Utica's attorney meant when he used the word, I don't know. It is obvious how I interpreted it.

Hindsight does not need corrective lenses; retrospect carries with it many speculations. Had I been told or remotely suspected that I was a target of the investigation, I would never have agreed to accept the presidency of the corporation if elected. You asked me if I was interested and, after some deliberation, I agreed to have my name placed in nomination. I knew at the time I was giving up a three-year contract at the bank and the opportunity for full vesting in the retirement plan. Nevertheless, I naively thought because I had acquired some knowledge of the corporation I could ably assist you and the corporation in carrying out its mission with very little, if any, loss of time, thereby contributing significantly to the economic strength and defense of the nation.

I am grateful for the position taken by the Board. It was unsolicited support and appreciated more than you will ever know.

Individuals are entitled to their own opinion. It is a tragedy when dedicated citizens (of which there are many) can't serve without coming under political siege. The Washington environment apparently affects people in many different ways. I regret I didn't understand more fully the vagaries of our government process as I never would have accepted.

Sincerely,

V. M. Thompson, Jr.

Overall Assessment

So, within the space of about six months, two presidents of the Corporation were driven from office, largely by the efforts of committees of Congress. Why were these attacks undertaken? The Cohen hearing was probably simply a favor to Monks in his power play. The Dingell hearing was part of a sustained attack designed to undermine the Corporation by forces that had been opposed to its creation. Dingell's behavior in the instance was in keeping with his reputation throughout Washington. A few years later, *The Wall Street Journal* commented in its lead editorial titled "Abuse of Power" (March 22, 1989):

> Washington witnessed a rare phenomenon this week: John Dingell tottered on his Energy and Commerce Committee throne. In the space of a few days, two of Mr. Dingell's victims achieved vindication. … While Mr. Dingell apologized, sort of, for his treatment of Mr. Gibbons, he also asserted that the power of congressional investigations makes the House of Representatives the 'grand jury to the nation.' Mr. Dingell's idea of what constitutes the legitimate use of state power suggests that Star Chamber may be closer to the reality. … Mr. Olson survived because of a spotless ethical record and because personal perseverance helped him fend off the inquisition.
> But such abuses of state power, whether by John Dingell or by Congress' Independent-Counsel cat's paw, will cease only when confronted by an equal force. The real solution to overreaching by Congress is political.

Be that as it may, let us return to the results of the attacks on the presidents of the Corporation. From a few shreds of appearances major ammunition to bring down the Corporation was fashioned. The totality of apparent missteps consisted of: (1) passing on the name of a reference for a fellow Board member, (2) the repayment of modest sums to the SFC when they were determined to exceed the intent of the relocation policy, (3) an assertion that there should have been more competition in awarding small contracts to individuals, (4) a conclusion that the Board should have been informed about an investigation concerning an outside institution, and (5) a conclusion that the ethics officer should have been told about an outside conversation that was in itself permissible. These were sufficient to bring down sitting officers, thereby leaving the aura of a scandal-ridden Corporation.

Appearance of SFC Extravagance

Other criticism leveled by the Corporation's enemies was that the Corporation was extravagant and that the staff and leadership were feathering their own nests, assertions that received prominence in the debate to terminate the Corporation, which is described in Chapter 13. Yet such criticism was totally without foundation: salaries and expenses were within the provisions of the Energy Security Act, and, if anything, lower than those of comparable federal agencies.

As a matter of record, the Board kept an eye on both government and the private sector. If in one matter, the private sector did it cheaper, that became the standard. In another, if the government's standards provided reduced costs, that became the measure. Let us examine the relevant factors: numbers of employees, salaries, fringe benefits, travel, quarters, and overall budget against the legislative allowances.

Number of Employees. Section 117(d) of the Act authorized the Corporation to employ up to 300 full-time professional employees to carry out the purposes of the Act, but no limitation was placed on numbers of non-exempt support staff. At its greatest size (in 1984), the Corporation employed about 230 individuals—of whom approximately 60 percent were professionals. Thus, it at no point employed much more than one-half the number of staff permitted by Congress.

Salaries. The Act authorized the Board of Directors to fix the salaries of individual officer positions and categories of other employees taking into consideration the rates of compensation in effect under the Executive Schedule and the General Schedule prescribed under law for government employees. If, however, the Board found it necessary to fix any of those salaries at a higher level (in order to hire qualified individuals) it could transmit its recommendations to the President, and if he did not disapprove within 30 days, such recommendations would go into effect. It was this provision, enacted by Congress itself that was used to paint the Corporation with the charge of profligate salaries.

Top Salaries. The high top salaries of corporate officers was one of the most politically sensitive issues. With that in mind, the Sawhill Board engaged a personnel firm, Towers, Perrin, Forster & Crosby, to determine what competitive salary levels would be for the Corporation's proposed officers. The results of the survey showed that all the vice presidents as well as the Chairman should be paid in excess of the Executive Schedule, with the highest salary of the Chairman set at $175,000. These recommendations were forwarded to the president, and, absent disapproval, went into effect. With the resignation of the Board a month later, and of all the officers except Leonard Axelrod (Vice President for Technology and Engineering) by spring of 1981, few of these salaries were collected.

When the new Noble Board was confirmed, the salary question was reconsidered for what was to be essentially a new slate of officers (excepting Axelrod). The new Board also engaged another personnel firm to make executive compensation recommendations. Following a review of these recommendations, the Board determined that only three salaries should be in excess of the Executive Schedule: that of the chairman, the vice president for technology and engineering, and the new position of the Corporation's president (who would be the chief operating officer of the Corporation). It concluded that the salaries for the chairman and the vice president for T&E should remain at the previously determined levels of $175,000 and $108,000 levels, respectively. Consequently, these salary levels did not require a new submission to the President. The new salary of $135,000 for the Corporation president did, however, require that the approval process be followed. Accordingly, the recommendation was transmitted to the president, and absent disapproval, went into effect.

In the event, Ed Noble, aware of the sensitivity of the pay issue in the Washington political arena, declined the approved salary level: for the next two years, he accepted only one dollar a year in salary. Thereafter, he only accepted $69,300, the top level allowed by the General Schedule for civil servants (lower than the top of the Executive Schedule governing top level political appointees). Ironically, he bore the brunt of the charges that the Corporation had voted themselves high salaries and were feathering their nests.

Thus, as a practical matter, only two employees received salaries in excess of that approved by Congress for other political appointees. Of these two, the president's salary of $135,000 was paid for less than two years, from December 1981 until August 1983, when Vic Schroeder resigned, and again for the one month that Vic Thompson held the position in 1984. During the period of greatest congressional criticism, only one salary exceeded the Executive Schedule.

Average Salaries. Overall salaries were in line with other comparable agencies. Comparability was established in the context of other agencies having a high proportion of staff and having little in the way of line operations. In response to the congressional criticism, the Corporation surveyed the average salary levels of such agencies as shown in Table XII-1.[148] The survey found that the Corporation's average General Service equivalent pay grade (excluding positions

COMPARATIVE AGENCY PAY GRADES

Agency	Average Grade and Step
Federal Mediation & Conciliation Service	12.6
General Accounting Office	11.5
Office of Management and Budget	11.4
Consumer Products and Safety Commission	11.2
NASA	11.1
Department of Transportation	10.9
Environmental Protection Agency	10.8
Federal Trade Commission	10.7
Federal Deposit Insurance Corporation	10.7

Table XII-1

equivalent to SES and executive level type positions) was pay grade 10, step 9. This was comparable to the pay scale of other agencies.

In addition, the picture for the Corporation's officers' salaries was not out of line. Officers' salaries in 1982, for example, averaged about $74,000. This was about $3,000 higher than the government's SES ceiling. But, SES positions were eligible for bonuses up to 20 percent of salary and 30 percent of SES incumbents received these. The Corporation's officers received no bonuses. Thus, on the average, the salaries were also comparable.

Fringe Benefits. The picture of frugality was clearer still when assessing the level of fringe benefits approved by the Board. Here, as elsewhere, the benefits consisted predominately of retirement and insurance benefits of various kinds, and was based on recommendations from George B. Buck Consulting Actuaries, Inc. as recommended in a report of May 27, 1982. The Corporation's total benefits package ran 23 percent of payroll, while the federal government's was estimated to run at 36 percent[149]. Moreover, given the government's unfunded liability for a pension system with inflation protection, the federal fringe benefits could run in excess of 80 percent of payroll.

The SFC's retirement plan consisted of three parts: Social Security contributions, a pension plan, and a savings plan. As a quasi-federal institution, all employees were required to participate in the Social Security System, and the Corporation made the necessary contributions, running at about 6.5 percent for the individual as well as the Corporation at this time. The SFC then made an additional payment amounting to about 5 percent of salary into a retirement plan. Finally, employees were permitted to place about 11 percent of the salaries into a tax sheltered savings plan and the SFC matched up to 3 percent of that amount, a 401(k) plan arrangement. The average cost to the Corporation of the non-Social Security component of the retirement plan was 9 percent of payroll.

The SFC retirement plan was what is termed a "defined contribution" plan in which all costs were fully funded as they were incurred. The federal plan, in contrast, is a "defined benefits" plan which carries a large unfunded liability. The SFC plan, for example, provided for no tenure, no cost of living increases, no "highest three years" base rate, no early retirement subsidies, and no insurance subsidies after retirement—all of which is included in the federal pension plan.

As a result, the SFC's plan was far less expensive than that provided by the federal government to civil servants. Nonetheless, congressional opponents attacked the plan on the basis that it vested more rapidly than the government pension (and most private pensions)—i.e., the SFC allowed an employee to be fully vested with regard to the Corporation's contributions after the first year of employment, whereas the federal government (at this time) did not allow vesting until the individual has worked a total of five years.

There were three responses to that line of attack. First, under the terms of the defined contribution plan, any forfeitures from someone leaving the Corporation before vesting were distributed among the other participants, so that longer vesting periods would benefit longer tenured employees, but would not save the government any money. Second, and most importantly, the Corporation was viewed by the vast majority of individuals as a short-term institution and not as a career opportunity. To have required a five-year vesting period would, in effect, be saying to most recruits that they would receive no retirement benefits by working at the Corporation. Moreover, the purpose of longer vesting periods at other institutions is to provide an incentive for employees to stay and be productive after an initial learning period. Obviously, there was little point for the Corporation to devise a retirement system with incentives for individuals to stay with the Corporation for longer periods. Finally, it should be noted that Congress itself was in the process of considering a basic reform for the federal retirement system so as to reduce its substantial costs. The direction of the reform was to go to a largely defined contribution system that was similar to the Corporation's, but more generous. That system embodied a one-year vesting period (eventually adopted by Congress for new government employees).

The other benefits provided by the SFC consisted mostly of health and life insurance coverage. The Corporation paid fully for an individual's health and dental insurance—requiring a contribution only for family participation. Similarly, the Corporation paid for group life insurance and disability insurance.

Yet, even with a relatively frugal plan of benefits, the Board kept a close watch on the total costs. When, after the first couple of years rising health insurance costs threatened to raise the total costs of benefits above 23 percent of payroll, the Board directed that the health insurance coverage be renegotiated to reduce coverage and, thus, premiums to the SFC. This was accomplished by requiring higher deductibles and by reducing slightly the percentage of claims that were covered.

Travel. Although travel costs were not a large fraction of the overall budget, they did represent a significant share of the variable discretionary costs. The program required a substantial amount of travel given that proposed projects were located from coast to coast. Due diligence called for all the sites of serious projects, pilot plants, and engineering offices be visited to confirm the existence of claimed levels of prior work. To be sure, the Corporation was somewhat more generous than the federal government in that travelers were not restricted to a per diem amount and the SFC paid for all reasonable lodging and incidental expenses, and only limited the total amount that could be claimed for food. Nonetheless, the Chairman personally kept an eye on travel activity, often challenging the need for specific trips, and no foreign travel (including Canada) could occur without his specific approval. Moreover, all travel was scheduled through a travel agency, which had instructions to schedule only with low cost airlines unless a higher-cost alternative were specifically authorized.

Because travel and travel-costs were to be brought up in critical congressional debates (see next chapter), their role deserves closer examination. Given that the SFC's mission at this time was to interest major investors in participating in sponsoring large costly plants, to review proposals for these plants, and to negotiate financial assistance agreements, a period of substantial travel was essential. The SFC reviewed proposals from about 150 projects throughout its existence. Any given solicitation might have thirty or so projects under review. To be sure, a number of these were not serious proposals, or at least represented projects of such dubious merit that they did not deserve in-depth evaluation:, and these were cut from the evaluation process early on by means of the maturity review. Nevertheless, the SFC routinely had about twenty projects that deserved careful scrutiny.

Most important were the technology and site reviews. There are as many fly-by-nights in synthetic fuels as in any other business. For example, one project that had submitted a proposal was the Consumer Solar Project. The sponsor hit on all the pop interest buttons—solar energy and non-polluting hydrogen to fuel automobiles. The chief executive was persuasive, even getting a lengthy favorable mention on ABC. It was all a scam. He dummied the data and was eventually imprisoned for postal fraud. Due diligence required that the SFC scrupulously avoid any possibility of funding such projects.

Indeed, due diligence, as undertaken by all serious investors, required seeing the pilot plants, reviewing their data, examining detailed design drawings, and relating them to the selected site. Doing so required travel, particularly by the Projects, Technology & Engineering, and the Environmental staffs but also by External Relations staff dealing with State and Local Governments, and the Inspector Generals staff.

Total Administrative Outlays. Section 120 of the Act authorized the Corporation's administrative expenses. Specifically, it authorized during any fiscal year: administrative expenses not to exceed $35,000,000, and additional expenses for generic studies and evaluations of individual projects not to exceed $10,000,000. Both of these amounts were specified as being in year 1979 dollars, which would be increased each year in line with the growth of the Gross National Product implicit price deflator. Over its existence, the SFC used only 21 percent of the funding made available to it.[150]

ROLE OF CONGRESS

When Congress created the SFC as a quasi-federal institution free of many administrative constraints imposed on federal agencies, it in no way relinquished any of its own institutional tools for oversight of the new entity. Indeed, it assigned numerous committees/subcommittees responsibilities for oversight of the Corporation as shown in the following list:

Congressional Oversight Committees

Senate

Committee on Energy and Natural Resources (E&NR)/ Subcommittee on Energy Research and Development (ER&D)

Committee on Governmental Affairs/ Subcommittee on Oversight of Government Management

Finance Committee

House of Representatives

Committee on Energy and Commerce/ Subcommittee on Fossil and Synthetic Fuels (F&SF)

Committee on Energy and Commerce / Subcommittee on Oversight and Investigations (O&I)

Committee on Government Operations/ Subcommittee on Environment, Energy and Natural Resources (E,E&NR)

Committee on Science and Technology/ Subcommittee on Energy Development and Applications (ED&A)

Appropriations Committee/ Subcommittee on Interior and Related Agencies

Committee on Banking, Finance, and Urban Affairs/ Subcommittee on Economic Stabilization (ES)

Committee on Appropriations/ Subcommittee on Department of the Interior and Related Agencies (I&RA)

Oversight committees have many tools at their disposal to carry out their responsibilities: public hearings, requests for investigations by the General Accounting Office (GAO) [now the Governmental Accountability Office], and official requests for information on all aspects of SFC operations. The above oversight committees carried out their functions professionally and with respect for the SFC. There was, however, ample opportunity for the opposition to the SFC to work at the Committee level to undermine its political support, notably by John Dingell (D- Michigan), chairman of the House Energy and Commerce Committee, and Mike Synar (D- Oklahoma), chairman of the House Committee on Government Operations/ Subcommittee on Environment, Energy and Natural Resources.[151]

Committee Hearings

Most hearings were routinely conducted by all of the committees holding jurisdiction over the SFC. For example, each year, the appropriation committees in both the House and the Senate would review the Corporation's budget, while the committees with technical oversight interest would periodically review the Corporation's progress in supporting projects.

The following is a representative list of both kinds of hearings Congress held in its oversight role (excepting the annual budget hearings) during the Corporation's existence:

Oversight Hearings

2/14/81, the House E,E&NR reviewed the Corporation's initial months to attack its hiring practices, salaries, and so forth, just after the SFC had lost the Sawhill Board.

4/3/81, the House ER&D held a progress review.

7/9/81, the House F&SF reviewed "Synthetic Fuels Policy".

7/27/81, the House ED&A reviewed "Synthetic Fuels Development".

9/17/81, the House E,E&NR held an "Oversight: Goals of the Reagan Board of Directors".

10/1/81, the House ED&A reviewed "Synthetic Fuels Environmental R&D"

4/2/82, the House F&SF held a general overview of the "SFC".

4/2/82, the House ED&A also held a hearing on "Socioeconomic Impacts of Synthetic Fuels".

6/9/82, the House E,E&NR reviewed "Synthetic Fuels Industry in Today's Economic Climate".

5/12/83, the House ES held a general progress review.

7/27/83 and 7/29/83, the Senate OGM held the hearing discussed above to review SFC management.

10/4/83, 10/5/83, 1/25/83 and 6/18/83, were a series of hearings held by the House F&FS on "Synthetic Fuels Policy".

10/18/83, the House E,E&NR reviewed the Great Plains

Coal Gasification Project's application for assistance.

12/8/83, the House E,E&NR held a hearing regarding "Examination of Procedures by which the SFC Selects Projects for Federal Financial Assistance".

4/3/84 and 6/27/84, the House O&I interrogated Vic Thompson and reviewed the aftermath.

5/16/84, the House E,E&NR held an "SFC Oversight" hearing.

6/6/84, 6/7/84, and 6/13/84, the House ED&A reviewed "The Status of Synthetic Fuels and Cost-Shared Energy R&D Facilities".

5/22/85, the House E,E&NR reviewed the proposed award of assistance to Great Plains.

Hearings, however, can also be used to achieve partisan ends, as occurred by those conducted by the "Government Operations Committee" chaired by Congressman Dingell in the house, and its Environment, Energy, and Natural Resources Subcommittee, chaired by Congressman Synar.

Previously, this chapter demonstrated how effective the hearing process was in destroying the reputations of the SFC's two presidents. These, however, were just a few among the ongoing use of these forums by the congressional opponents of the Corporation.[152] Two additional hearings dealing with the Santa Rosa and the Great Plains projects are especially illustrative at how ad hoc hearings can be employed for political ends.

The Hearing on Santa Rosa

To refresh the reader's recollection (from Chapter 7), Santa Rosa was a tar sands project that had been proposed by the Solv-Ex Corporation that was to use a proprietary process they had developed to extract petroleum from tar sands by mining the sands and using solvents to separate the oil from the sand. Solv-Ex developed the process, built and operated a pilot plant, and designed a commercial production facility. The Corporation considered it a promising project because it would develop a new technology applicable to tar sand deposits and thereby add to the diversity of technologies available to the country.

Accordingly, the SFC issued the project a letter of intent, but a letter contingent on resolving several points. Most were straight forward, but two would require investment by the project sponsors: i.e., the characteristics of the bitumen to be produced by the project had to be verified to the satisfaction of the SFC; and the Partners or affiliates had to have operated an integrated pilot plant for at least 90 days (including a run of at least 21 consecutive days), thereby confirming the technical and financial feasibility of the project.

The SFC was exercising due diligence to ensure that the resource base was both sizeable enough and appropriate to the technology before federal monies were committed. By signing the letter of intent, the Corporation provided the private sponsors with the incentive to make the financial investments that were necessary to reduce the uncertainties (at no cost to the Government).

The uncertainties in question were largely twofold. First, the technology had previously been run only in a batch mode and problems typical in solids handling processes suggested that a demonstration of the eventual continuous mode of production would be prudent. Secondly, the Corporation had already engaged the Laramie Energy Technology Center to review the proposed resource at the site to confirm its suitability as a feedstock. Laramie's report was basically confirmatory, but it indicated that some additional checking would further reduce uncertainty.

As events transpired, the project sponsors made the investments in beginning the mine and running the feedstock through the pilot plant in a continuous mode, but discovered that the sands in the proposed site did not have the characteristics anticipated. Discontinuities in the richness of the tar in the sand reduced the overall oil content by a crucial couple of percent, enough to undermine viable economics.

Consequently, Foster-Wheeler withdrew from the project and the Corporation cancelled the letter of intent (since a key condition could not be met). That should have been the end of the matter. The system worked as intended and no federal funds were expended. The House of Representatives Environment, Energy, and Natural Resources Subcommittee of the Committee on Government Operations thought otherwise and scheduled a hearing that was held on December 8, 1983.[153]

Appendix E reviews the main questions pursued by the Committee: was there a flaw in the SFC's evaluation process; were new equity investors mislead by the letter of intent; and did the SFC fail to get adequate input from local groups and agencies regarding the project's possible environmental impact?

The lengthy hearing showed that the SFC evaluation process worked as intended—i.e., it identified uncertainties in the proposed project plan and outlined measures required to resolve those uncertainties, which were undertaken at no cost to the government. In addition, there was no evidence that any investors were mislead, and detailed testimony demonstrated that there was ample consultation with all parties and agencies potentially concerned with the environmental impact of the project.

This positive testimony did not present an obstacle to a mostly hostile Committee, which issued a report containing the following:

> The SFC's pool of financial resources, although large, is nonetheless limited. Accordingly, the Committee is concerned that if flawed staff evaluations result in expenditures of significant time and resources on proposals which ultimately prove fruitless, other perhaps more deserving projects may go unaided as a result. (In fact, the hearings failed to disclose a single instance of where a deserving project had somehow failed to get adequate staff attention because of time spent on Santa Rosa).
>
> SFC staff failed to verify the accuracy of some of the information submitted by the sponsors of the Santa Rosa Tar Sands Project. (The Subcommittee review produced documentation contradictory to the staff conclusions provided to the Board.)

The citation does not specify what information was inaccurate, but it presumably applies to statements by Getty Oil and Amoco that they were not interested in using Solv-Ex's technology. Synar tried to conclude that the technology was not replicable and that the SFC staff had not ferreted out this information. During the testimony, however, Mr. Axelrod noted that these firms were developing technologies of their own and had other competing interests.

Moreover, the firms did not indicate that the technology was not applicable, but, only that they were not interested.

> The New Mexico State Parks Division was not consulted by the SFC about the impacts of the Santa Rosa Project, although mining activities would have incorporated portions of an existing State Park.

This point had been refuted during the hearing.

> The Committee recommends: (a) establishment of an intra-SFC committee, to be represented by members of the project evaluation teams, for the purposes of re-evaluating the SFC's project selection procedures. ... the SFC repeatedly has been forced to waive established selection criteria in its efforts to advance project award decisions. ... (b) The Committee recommends that the Corporation maintain firm selection criteria to ensure levels of project development that are necessary for informed Board decisions. ... The Committee recommends that the SFC reassess staff evaluations of all eight projects with pending or proposed letters of intent ... .The Committee recommends that: (a) The SFC improve its public notification procedures which accompany the project advancement decisions by the Board. ... The SFC should be more cognizant of the possible shortcomings of federal and State regulations in protecting the environment and public health with regard to SFC projects."... Because of information uncovered during the course of the Subcommittee's review, the Committee is concerned that infractions of securities law may have occurred. ... Accordingly, in order to clear up the matter, the Committee is referring its information to the Securities and Exchange Commission ...

In fact, the Committee had not uncovered any evidence that would have justified these platitudinous findings. Nonetheless, the Corporation did set up independent evaluation teams and redid all the evaluations per the Report's recommendations.

The Report was too much for the entire Subcommittee to swallow, and Congressmen Dan Schaefer, Robert Walker, and Thomas Kindness issued dissenting views:

> [With regard to the SEC referral, any] fair and reasonable review of the record does not readily substantiate such a finding. To the contrary, the record shows that the actions of Solv-Ex and Morgan Stanley resulting from the 1983 resource evaluation report and mining site activities were proper ... . In addition, whether or not one agrees with the conclusion that an error by a SFC member resulted in the SFC board getting incomplete information regarding the Santa Rosa Tar Sands Project, we wish to stress that the project did not receive any financial assistance from the SFC. Thus, it is fair to conclude that the letter of intent process for screening projects did work. ... The action of this Committee can only discourage and impede the participation of small high technology companies in government sponsored ventures. This result is obviously undesirable since it should be clear legislative and public policy to foster development of Small Business with new and revolutionary technologies.

The Hearing on Great Plains

Congressman Synar's Subcommittee held another project-specific hearing on May 22, 1985, regarding the Great Plains Coal Gasification Project.[154] This hearing was notable in that it was the only example where Congress employed the hearing process to undermine ongoing negotiations with a project.

The Corporation had just negotiated the terms of assistance for the Great Plains Project pursuant to the competitive Solicitation for Coal or Lignite Gasification Projects. The project had already received assistance from the Department of Energy in the form of a $2.02 billion loan guarantee made in 1981. As discussed in an earlier chapter, the loan guarantee was sufficient to raise the funding necessary to construct the project, but it did not provide any revenue protection in the event of falling energy prices (all the commitments negotiated by the Corporation included price guarantees). When the weakening energy price picture became clear to the project's sponsors, they applied to the Corporation for price guarantee assistance. The term sheet that emerged from negotiations would have provided the project up to $820 million in price guarantee payments, in exchange for which the sponsors would have required to put up an additional $190 million in equity toward paying off the government guaranteed loan, and to accelerate the payment of that loan through the project's cash flow.

The government's potential for a loss in covering the $1.5 billion loan (the amount actually drawn down of the $2.02 billion made available) would be virtually eliminated by making the $820 million price guarantee. When factoring in the time value of money, however, the options were more nearly equal in financial terms. But the new arrangement virtually assured the long-term continuous operation of the plant and the benefit of continued technology development. All in all, it was a reasonable deal for all involved, and it was clearly within the authorities of the SFC.

So, on what grounds was the hearing held? First, Chairman Synar mischaracterized the nature of the proposed deal by making it sound as though the project had already received the one and a half billion dollars of taxpayer assistance and would be receiving an additional $820 million of assistance in the form of price guarantees. His opening statement went along the following lines: "Today the Subcommittee … will review a proposal by the U.S. Synthetic Fuels Corporation to award an additional $820 million in Federal assistance to the Great Plains Gasification Project in Beulah, ND. … To date, the project's sponsors have borrowed almost $1.5 billion of taxpayers' money. … The issue at hand today is not whether or not we shall have Great Plains but, rather, the manner in which we shall have Great Plains. The Project is built. The taxpayers' money is spent. Yet, despite the well-managed construction and encouraging technical operations, this project has been, and continues to be, a financial nightmare. It is simply not economic under anyone's expectations. The bottom line is that the sponsors cannot operate Great Plains and service their debt to the payers of $1.46 billion. To get their money back, we are now told that we must spend more taxpayer dollars."

The subcommittee had an analysis of the proposed assistance done by the U.S. General Accounting Office and had members of that staff in to testify. While this testimony added considerable detail, it did not change the facts in hand. Even though the amounts of assistance and its form can be second-guessed, as the SFC's vice president for finance told the Subcommittee: "This is a negotiated transaction and it can only be concluded when all the parties are willing to accept it. The parties have various motivating factors and all of these have to be dealt with to some greater or less degree. But the purpose of the deal is to have the plant operate. It achieves other things that are important to the parties, and that's the way a negotiated deal comes to a settlement."

The second main attack by the Subcommittee was more awkward for the Corporation. It dealt with the ability of the Board to transact business with regard to Great Plains and whether the Board had a working quorum to approve a deal. The situation was complex, but can be summarized along the following lines. For the Board to conduct business, the Energy Security Act required that at least four members of the Board had to vote in the affirmative, regardless of the number of available Board members. And the reader may recall that the White House, despite the apparent agreement of the prior year, had only nominated three additional members to the Board rather than five to bring it to full strength. Thus, four of the five available members had to vote in the affirmative for a motion to pass.

Moreover, Subsection 118(c) of the Energy Security Act prohibited any director from voting on any matter in which, to his knowledge, the director had a financial interest. For this purpose, he must also consider his spouse's interests. The Act, however, provided that the prohibition may be waived by the Board of Directors if a majority of the Board determines that the financial interest is too remote or inconsequential to affect the integrity of a Director's services as to the matter. This latter provision added a particularly intriguing twist to this particular matter. For example, if two of five directors have inconsequential holdings, can either of them vote to determine that the other's holding is inconsequential and not be perceived as voting on a matter in which they have an interest (since it will not at that point have been waived)?

Unfortunately, that exact situation obtained with regard to the Great Plains vote. Specifically, Paul MacAvoy sat on the Board of

Combustion Engineering, Inc. and held its stock. Combustion Engineering was not an owner of Great Plains, but had done design work and engineering and still had some open claims. The SFC ethics officer determined that this was a financial interest (though not necessarily consequential).

At the same time, Eric Reichl's wife held 1,333 shares of Midcon stock. Midcon was a 15 percent owner of Great Plains. This investment was apparently a small part of her holdings and Reichl was quoted as saying "It's inconsequential … . As a matter of principle I don't think I should have to sell it."

An opinion provided to the Sub-committee by the Congressional Research Service concluded that neither Board member could legally vote on the consequentiality of the other's holdings. The Subcommittee spent considerable time on this issue, but did not advance the matter. For all practical purposes, if there had not been a legal resolution of the matter, the SFC staff felt that Mrs. Reichl would probably just have sold the stock if the crucial vote were imminent. However, such a vote never came to pass.

The Subcommittee also heard testimony from the Great Plains Project and from the Department of Energy, but no significant other issues were raised. While the hearing revealed no new information, it seemed to shed negative light on the negotiated deal and set the stage for the Secretary of Energy to withdraw support for the agreement, effectively killing it (see the discussion in Chapter 11).

Requests for Information

In addition to congressional hearings, there were three widely utilized tools whereby congressmen could oversee the Corporation's efforts to achieve the mandate Congress had established in the Energy Security Act: briefings by SFC staff, formal requests for information, and investigations by the General Accounting Office.

With regard to briefings, the SFC routinely met with congressional staffers on all ten oversight committees, especially after significant board actions. During the active years, this could amount to thirty or more briefings a year. These provided the staffers with the background for board actions, an opportunity to more fully understand their significance, and to gain insights into how the SFC was addressing problems early on in finding suitable projects, capable sponsors, and private equity.

While much of this routinely allowed staffers to track progress, congressmen suspicious of the Corporation could also use these tools to investigate perceived problems or shortfalls.

While oversight committees' requests for information routinely track progress and illuminate policy issues, these inquiries are also a useful means for Congress to have organizations defend their actions and procedures on the public record. It is altogether salutary for organizations spending public funds to be held so accountable.

Yet such measures can serve other, arguably more partisan, purposes as well. The following discussion highlights the nature of information requests made of the SFC in terms of method and substantive interest, illustrating the problematic as well as the positive. Regarding the former, one sees how Congressman Dingell, who had boasted of his intention to become the nightmare of the Corporation, was able to probe SFC operations. Even though batteries of official letters and investigations were launched, these produced no evidence of the Corporation failing any of its duties under the Energy Security Act at the end of the day. Nonetheless, their number and their tendentiousness served to cast the SFC in a negative light during the years of ebbing political support.

The targets of most of these information requests were: the legislative basis for issuing letters of intent, the "secretiveness" of the SFC vis-à-vis sponsor-provided information, and whether the Corporation was abiding by its published procedures in the context of competitive solicitations. To be sure, all of these are potentially important issues that Congress might want to have on the record. Yet given the repetitive nature of the same questions and an apparent lack of interest exhibited by the committee' members in the specific answers that Corporation staffers laboriously provided them, suggested that other motivations were at work. A few examples make this point.

As the reader may recall, letters of intent were an innovative way the SFC employed to provide incentives for project sponsors to gather equity and to make projects more likely to come to fruition, without committing any Corporation funding. Dingell's letters requested information on the basis for such instruments, particularly after the first one had been signed with the First Colony Project: in December 1982, he fired off a lengthy set of questions. After the SFC replied with a hundred or so pages, he followed with another batch of twenty-five questions covering much of the same territory on July 18, 1983 (see Appendix F). After the Corporation responded at length to these questions, Congressmen Dingell and Broyhill (the senior Republican member of the Committee at the time) sent still another letter to the Corporation regarding the First Colony Project on November 18, 1983. This letter raised the same general questions regarding the use of letters of intent, but added questions related to how the Board dealt with dissenting staff views, on the replication potential of peat, on pilot testing of peat, on environmental concerns, and on specific financial aspects of the proposed financial assistance.

The SFC answered that letter on January 10, 1984, just to receive still another from the two congressmen on February 16, 1984.[155] This letter followed up on the prior SFC response to continue to challenge the use of letters of intent, while it raised some new issues: was the SFC a secretive organization seeking to evade oversight by shredding documents and using "safe houses" and did the Corporation relax solicitation standards when advancing projects through the review process?

The basis for the use of letters of intent was covered in an earlier chapter (Chapter 7) and will not be repeated here. But since the charge that the Corporation was inappropriately secretive was one repeated in Congressional speeches with some frequency (see the next chapter), it is worthy of a brief review here. The SFC had a policy of shredding multiple copies of sensitive documents when they were no longer required by review teams. The solicitations required sponsors to provide multiple copies of documents to facilitate staff review by teams generally having about six members. These materials often contained proprietary and business confidential information, which the Act charged the Corporation with protecting, subject to penalties. The SFC policy was to destroy copies when no longer needed, keeping only one for the Central Project files. The only sure method of safeguarding these sensitive materials when discarding them was to shred them (common practice in the defense establishment).

The Safe House process was designed to give the SFC staff access to highly proprietary technical data, from which the project's claims could be verified. These materials went far beyond what would be needed in the Corporation's permanent files and would have only caused the sponsors' great unease, had they had to release copies. Consequently, the sponsors made such materials available to the Corporation only in Safe Houses, but for however long the evaluation process continued. This approach was consonant with industry as well as DOE practice to preserve confidentiality in new technology ventures. In some instances involving venture capital, such materials are made available only to a third-party expert who has signed extensive confidentiality agreements.[156]

Finally, with regard to the so-called waiving of requirements, the Corporation never waived a requirement stipulated in a solicitation. What the SFC did do was pass a project through an evaluation phase subject to completing a requirement by a time certain. For example, if a project had to reach a Level II/III design and one system or another still required the completion of some engineering work, the Board might allow the project to move into negotiations while completing the work. In no instance was an award made unless all requirements had been met.

The Corporation replied to the letter raising all of these issues on March 28, 1984. The reply was one hundred pages in length (the same month that the Corporation prepared and a sent a response to the Senate Energy Committee, also approximately one hundred pages in length).

While all the Oversight Committees sent such letters to the Corporation, most did so sparingly on issues of importance to policy and for the purposes of establishing a record. By and large, it was only the Dingell Committee and the Synar subcommittee that used the letters in a harassing way. Indicators that these letters were not intended to serve a constructive oversight purpose were that the same questions would be re-posed over time even after answers had been supplied, and the congressmen would make repeated factual errors in hearings and in congressional debate (see the next chapter) that had been flatly contradicted by the SFC's responses to his enquiries.

In the end, no constructive finding ever emerged from the Committees' reviews of the information submitted. Indeed, there was never an instance found in which the SFC was not operating in accordance with the Energy Security Act and with the law in general.

Investigations

Another traditional oversight tool entails investigations by the U.S. General Accounting Office, an arm of Congress. According to the log kept by the Corporation's Office of the Inspector General (responsible for coordinating the SFC response to investigations), the GAO was tasked with 34 inquiries over the Corporation's existence, of which fully fourteen were initiated by Dingell's committee or one of its subcommittees. These imposed a considerable workload on the SFC staff, given that one or another GAO team was in residence for most of the Corporation's existence. There were often several investigations proceeding simultaneously. Not all of these investigations resulted in a formal report (presumably because nothing of note was uncovered to justify the effort). On at least one occasion, the report was quashed because it did not produce the desired result: Dingell had the GAO investigate the legality of the Amendment Agreement reached between the Corporation and Unocal, and when GAO concluded that the Agreement was legal and binding, they were apparently urged not to issue a report.

The following is a list of reports issued by the GAO through early 1985 (the log shown by the IG might not have caught other reports issued during the waning days of the Corporation). Shown below are the dates of issuance, the title, and the requestor for each of the reports:

GAO Reports on the SFC

12/8/80 - "Special Care Needed in Selecting Projects for the Alternative Fuels Program" (Dingell)

7/10/81 - "Synthetic Fuels Corporation's Management of Demonstration Projects Would be Limited" (Domenici)

8/5/81 - "The United States Synthetic Fuels Corporation's Project Selection Guidelines Need Clarification"

10/18/82 - "Evaluation of Administrative Procedures at he Synthetic Fuels Corporation" (GAO Internal)

10/22/82 - "Environmental and Socioeconomic Status of the Hampshire Energy Project" (Moffet)

2/2/83 - "Synthetic Fuels Corporation's Use of an Expert Panel and a Staff Assistance Agreement" (McClure)

4/12/83 - "Review of the United States Synthetic Fuels Corporation's Financial Statements for the Year Ended September 30, 1982" (GAO Internal)

11/20/83 - "Circumstances Surrounding the First Colony Peat-to-Methanol Project" (Dingell)

1/19/84 - "Letter from the Comptroller General of the United States to the Honorable Tom Corcoran re the legality of the United States Synthetic Fuels Corporation providing price guarantees to Great Plains Gasification Associates" (Corcoran)

2/7/84 - "The U.S. Synthetic Fuels Corporation's Contracting with Individual Consultants" (Dingell)

2/15/84 - "Letter from the Comptroller General of the United States to the Honorable John D. Dingell re whether the United States Synthetic Fuels Corporation is adhering to the restrictions on financial assistance placed on it by the Energy Security Act" (Dingell)

3/9/84 - "Federal Efforts to Control the Environmental and Health Effects of Synthetic Fuels Development" (Sharp)

7/11/84 - "The Synthetic Fuels Corporation's Progress in Aiding Synthetic Fuels Development" (Sharp)

7/23/84 - "Plans to Award Additional Financial Assistance to the Union Oil Company Oil Shale Program" (Synar)

9/20/84 - "Procedures Need Strengthening in the U.S. Synthetic Fuels Corporation's Conflict of Interest Program" (Dingell)

2/21/85 - "Financial Status of the Great Plains Coal Gasification Project" (Synar).

7/10/85 - "Synthetic Fuels Corporation's Profit-Sharing Provisions with Six Proposed Projects" (Synar)

8/15/85 - "U.S. Synthetic Fuels Corporation's Contracting Policies and Practices for Consulting Services" (Synar)

Some of the titles suggest more significant findings than were in reality the case as seen by examining a couple of the reports, beginning with #8. Dingell, in trying to undermine the First Colony Project had asked GAO to examine a series of issues including "the legal authority for such a letter of intent" and "the reasons why the Corporation staff recommended against federal backing of the project." GAO answered each of Dingell's questions. Notably, in the two instances cited, the answers were "Nothing legally prohibits the Corporation from being a party to a letter of intent" and "At the time of the signing of the letter of intent, the Corporation staff was not recommending against federal backing of the project." The other answers were simply factual replies. The report did have two recommendations for the Board of Directors: "either formally authorize or prohibit the chairman's signing of any future letters of intent, and assess the adequacy of Corporation information on two key questions regarding the peat-to-methanol project—replication potential ... and successful testing of the North Carolina peat in a pilot-scale facility ... to assure itself that both have been fully resolved." Both were done.

In another example, GAO's recommendations in #10 above were to have the Corporation "follow the Board's policy and award individual consultant contracts on a noncompetitive basis only after it determines that unique expertise needs and/or a time-critical situation makes competition infeasible; provide a written justification demonstrating that competition is not feasible for contracts costing $25,000 or more; and follow its guidelines by (1) comparing consultants' charges with the cost of hiring permanent employees before awarding contracts to individual consultants, recognizing that for some contracts the documentation may be brief and (2) monitoring consultants' performance over the life of the contracts."

These recommendations resulted in establishing formal procedures for what had already been standard practice. Moreover, although the Act clearly excluded the Corporation from the burdens of the Administrative Procedures Act, GAO had the example of the rest of the federal government in mind with these recommendations. The reasoning underlying their recommendation was that since the Corporation's procurement procedures did not explicitly exclude small contracts with individuals from competitive require-

290 The Saga of the U.S. Synthetic Fuel Corporation

ments, they should be revised to do so. GAO did not question the reality that it is not cost effective to carry out a competition to engage someone for a brief service.

The Board approved changing the procedures to allow such exclusion. Afterwards, the Corporation compared its actions with other federal agencies. It found that all agencies engaged consultants on the basis used by the SFC. Moreover, in relative terms, the SFC had far fewer non-competitive procurements and for lesser amounts. Another GAO other recommendation went to correcting inadequate documentation in some of the files, i.e., requiring a written analysis to show that it makes more sense to hire an individual for a few months on a contractual basis rather than to hire a person full-time.

In still another example, #17 above, the GAO report merely made its own calculations of profit-sharing revenue projections and compared the results with the estimates and the methodology employed by the Corporation, making no recommendation.

These examples are fully representative. For all the significant effort expended by a highly professional GAO staff, no instances were uncovered in which the Corporation was not acting within its legal authorities. Moreover, however well meant, none of the recommendations saved the government money or materially improved the Corporation's management.

At the end of the day, four years of hearings, GAO investigations, and formal requests for information were ample to create the impression of a "scandal-ridden" SFC, taking a substantial toll on political support for the Corporation. Chapter 13 will describe how these efforts were brought to bear in the congressional debates that led to the termination of the Corporation. But this chapter will conclude with a review of how the media contributed to creating a negative image of the SFC.

MEDIA COVERAGE

The media did little to enlighten readers about the Corporation's accomplishments or to provide balanced coverage of the political debate. This outcome was not due to lack of effort on the part of the Corporation to keep them informed. Indeed, the SFC held press conferences after every Board meeting at which significant action had been taken.[157] At those conferences, materials were made available to describe the actions as well as the context in which they had been taken. Moreover, the briefings tended to be well attended by the energy trade press, mainstream newspapers, congressional staffers, and, of course, the environmentalists.

To be sure, major newspapers and the energy press reported the basic facts of Board actions in a reasonably straightforward fashion, generally in the business section. For example, if the Board passed certain projects through a procedural milestone such as the strength review, or the Board approved the awarding of financial assistance to a certain project, the facts would get reported without distortion.

Nonetheless, if one looked only at editorials, a few columnists, a televised news special, and much of the context of straight news reports, the picture presented was very different. This is particularly true regarding media considered influential within the Beltway. Without exception, all of these reflected the narrative promoted by the Corporation's opponents without any attempt at independent verification or balance.

The Washington Post

The *Washington Post's* coverage of the Corporation was supportive when it was created by a Democratic administration and Congress, but cooled after a Republican administration came to power and appointed a new Board. In any event, the *Post*, which could be considered the newspaper of record on events in the nation's capital, hardly conveyed what was going on with the Corporation, what the various forces were after, or what was at stake. If one goes through the *Post's* yearly indexes, there are roughly ten entries per year. Of these, most contained some factual reports of Board actions with regard to the projects under consideration, but reporters consistently introduced a negative spin.

Three examples are representative of the entirety. First, during this period the *Post*, generally antagonistic to CIA Director Bill Casey regularly sought targets of opportunity to embarrass the Director. One of these occurred on March 31, 1982, after the Board had just passed five projects from the Initial Solicitation through the Strength Review, when the *Post* printed an account on the business page. A piece reporting on the first five projects being seriously considered would certainly be newsworthy. This article was not. The headline was "Casey Tied to Energy Firm's Bid for U.S. Backing." (*The Washington Post*, March 31, 1982). The firm, the First Colony Project, was being supported and financed by a number of substantial groups including Transco, Koppers, and J.B. Sunderland (the owner of the peat reserves). They were providing the capital and would get virtually all of the return if the project were successful. One of the sponsoring groups, however, was the Energy Transition Corporation, which did much of the work in preparing the proposal and doing the environmental analysis. This Corporation had five partners, one of whom was Bill Casey, although he was inactive and had no role whatsoever with regard to the project. That group would not have received any return from the project until years after it was built and clearly profitable. The bulk of that would have gone to the active partners. The net present value of any return that Casey could have realized was clearly negligible. While the article eventually stated that there was no impropriety, the Casey angle represented a significant element of the *Post's* reporting. And what was it doing in the headline?

Second was an article published that same year with the title "Suspect Firm a Finalist for Synfuels Aid" (*The Washington Post*, January 1982). The lead paragraph of the article read, "Consumer Solar Power Corp., a California company that claims to make fuel out of water and whose top executives have been accused of fraud by two federal agencies, is one of the finalists for the $8.4 billion in federal subsidies from the U.S. Synthetic Fuels Corp." The article went on to report the standard professions of innocence by the principals of Consumer Solar, and statements by the SFC that it could not prejudge the case. Not a happy picture. But the thrust of the article was nonsense. Consumer Solar never came close to being considered for an award by the SFC. It never passed the most elementary of maturity reviews, much less a strength review. At the time that the article was written, 27 of the projects that had applied under the initial solicitation supplied the supplementary data that had been required by the new Board under Chairman Noble before the SFC would begin to evaluate projects. The Board had taken absolutely no action with regard to any of the projects except to note which of them had met the deadline for submitting information.

Indeed, at the first opportunity (the maturity review), the Board eliminated the Consumer Solar Project from consideration. But when Bill Rhatican, the SFC vice president for external relations was asked for a comment, he could merely say that the SFC's deliberations were not yet at a stage where Consumer Solar's legal difficulties were a determining factor. After all, Consumer Solar was not to be deprived of due process under the SFC's published procedures for consideration. And the *Post* had ample means of ascertaining those facts.

A third example of tendentious reporting by the *Post* was an article published on July 24, 1984, again in the business section. The headline was "Synfuels Guarantee Put at $92 a Barrel" (*The Washington Post*, July 24, 1984). The headline and the article were derived from a press release issued by Mike Synar's Government Operations Subcommittee on the Environment on the previous day. That press release was putting Synar's spin on a GAO Study, which Synar had requested, but which had not yet been issued. The *Post*, published this account from the press release without waiting for the relatively factual report itself, giving plenty of play to Synar's characterization: "decrying the synthetic fuels tentative contract as 'corporate welfare'" and "It is inconceivable that while we are holding the line on lunches for school children, assistance for the needy and health care for the elderly, we are guaranteeing Union Oil Co. federal handouts of roughly $1 million a day."

What was wrong with this picture? The principal disingenuous element was that the GAO report itself gave no prominence to the $92 a barrel and the $92 figure was totally misleading. The GAO report had to do with the letter of intent that the SFC had signed with Union Oil for the Phase I expansion of its Parachute Creek facility. The potential assistance would have provided price guarantees of $60 a barrel (or $67 per barrel if the new Unishale C technology were employed in the larger Phase II facility) adjusted for inflation. The GAO report noted that if there were years of inflation at the level of five percent per annum, then at some time in the future the guaranteed price would be $92 per barrel. But, of course, no one knows what a future level of inflation will actually be. The point of the Synar press release was that every reader would be comparing the $92 figure with other current prices rather than the correct $60 (or $67) figure. More importantly, as noted elsewhere, the per barrel price guarantee figure needs context. First, the SFC would only be paying the difference between that figure and the market price. In addition, under the contract (based on the law), the guarantees were only for ten years whereas the plant would have a thirty-year life. In effect, the billions invested by Unocal would have to be recovered in the ten year period. The SFC calculated that if one allocated government assistance over the total amount of production anticipated from the facility in its lifetime, the subsidy per barrel would only amount to one or two dollars.

Once again, the *Post* could have obtained the facts, could have waited for the full GAO Report, or it could have checked with the SFC. But it was already in the mode of allying itself with the Corporation's opponents. Indeed, why should the story have gotten *any* prominence? At this time, the Corporation had eight letters of intent outstanding, but no quorum of the Board to follow through on them. In the event, this one was not consummated (though Unocal did eventually receive an amendment to the existing contract). The point was that Congress was at that time debating whether to cut the funding for the Corporation, Synar was a major antagonist to the Corporation, he was looking for negative headlines, and the *Post* was willing to lend a hand—in the news section.

Otherwise, during a several-year period, the *Post* had about twenty or so articles dealing with the Corporation, most of which were short and to the point, dealing with the numbers associated with Board actions[158].

The Wall Street Journal

The Wall Street Journal appealed to the right rather than the left of the political spectrum. Its performance was more in accordance with journalistic precepts: objective in the reporting and opinionated on the editorial page. Its accounts on actions by the SFC were brief and usually on interior pages.

Over the years, it produced several editorials attacking the rationale of the synthetic fuels program, which was not surprising given the philosophy of the newspaper. In effect, however, the SFC received journalistic support from neither the Left nor the Right of the political spectrum.

In one key instance, however, the *Journal's* approach became misleadingly partisan. During 1984 and 1985, when the SFC's funding and then its existence were under debate, they gave prominent exposure to a meretricious piece by John Herrington, the Secretary of Energy, but refused to print any rebuttal by the Chairman of the SFC or congressmen to balance the record.

Specifically, Herrington's op-ed was published just after he had reneged on his Department's agreement to support a modified assistance package to the Great Plains Project, and when the Board was considering assistance agreements for both the Parachute Creek and the Seep Ridge projects. His contentions were bafflingly ill informed for a Secretary of Energy as seen by his major assertions in a piece entitled "The Synfuels Energy Dinosaur" (*The Wall Street Journal*, October 9, 1985).

Early on, he states that $900 million was being considered for Parachute Creek. Actually, the Augmented Agreement would have provided $500 million (above the $400 million that DOE had approved four years earlier). Then, in several places he states "Parachute Creek has major technological weaknesses. ... The Parachute Creek project still had not operated successfully beyond a period of a few days, despite the investment of $800 million by the company over the past six years. ... Parachute Creek's scraper system—a mechanism that ejects spent shale—failed and still hasn't continuously worked as advertised. When it does work, the spent shale is coming out at a too high temperature (900 degrees Fahrenheit)." For a man charged with developing alternative energy technologies for the country, this was an incredible set of statements.

A brief rejoinder would note, first, that the nature of new technology development is that lengthy time is usually required to get scaled-up facilities to operate. Earlier this chapter noted that SASOL in South Africa required three years after construction to get to about 75 percent of design output despite building on the technology that the Germans successfully used during the war.

Second, there was nothing wrong with Parachute Creek's scraper system, it always functioned. The problem arose with handling the spent shale after it left the retort. It is true that the handling problem was connected with the high temperature of the spent shale, but his implication that the 900 degrees was too high was ludicrous. The kerogen will not flow from the crushed shale unless the temperature gets up to at least that temperature. The problem is how to handle the flow and wetting of the hot spent shale at that point.

Finally, in any event, Parachute Creek learned how to deal with the hot spent shale the following year achieving 35 percent of design output. By 1989, they achieved 70 percent of design levels. This pattern of hard won achievement was pretty much along the lines predicted by the Rand study of new technology discussed in earlier chapters.

Herrington also stated, "Both Parachute Creek and Seep Ridge have technologies applicable to only a small portion of U.S. oil shale resources. ... neither would make a lasting contribution to U.S. energy security." To the contrary, Parachute Creek technology is applicable to all shale resources. Those resources (the richer, more economic resources) contained oil the equivalent of 600 billion barrels—twice the amount of oil in the Middle East. Without proving the Parachute Creek technology, the nation would have absolutely no proven technology to access that oil in the event of need!

Further, he argued, "As a result of these fundamental changes in the energy market place [the failure of the price of oil to rise as projected], virtually no projects pending before the SFC are likely to become economical in the foreseeable future. ... There is little point in building demonstration projects when the fuel costs are two, three or four times that of current and anticipated market prices." As discussed earlier, however, no first-of-a-kind facility was likely to be economic. That was the rationale for government involvement. The question is whether the plants were an essential step in the direction of developing technologies that would be economical. Moreover, given the long lead times of a decade or longer, it was essential to be developing the commercial scale technologies when they were not economic to be sure to have them when the price of oil when up.

Next, and most misleading, he stated, "There is no compelling reason to subsidize construction of model synfuels plants when advanced technology, now under development by industry and the Department of Energy, will ultimately surpass these plants' existing technologies." The facts at the time were that there were no technologies being developed by the Department. While industry was working to a limited degree at the pilot scale, it had made clear that it would not undertake the risks of commercial scale-up without government assistance.

Considering that the Secretary of Energy is charged with leading the nation in matters having to do with energy security and technology development, one is at a loss to account for an article of this nature at a time of important congressional debate. Given that he lacked an energy or technical background (he came from a personnel position in the White House), it was probably not surprising that he adopted an ideological perspective derived from the conservative side of the Reagan administration, which was strongly opposed to Government involvement in the private sector.

The *Journal* not only refused the requests by the Corporation for a chance to correct these extensive misrepresentations,[159] but also one from Congressman Michael L. Strang from Colorado who wrote the *Journal* a rebuttal incorporating most of the points mentioned above.[160]

Other Written Media

Other print media paid even less attention to synthetic fuels. Of course, local newspapers in communities with pending projects tended to be overwhelmingly in favor of the SFC and its efforts, e.g., all of the Colorado newspapers supported shale oil development. Otherwise, over time the impact of congressional attacks began leaking into editorials with increasing references to a troubled corporation. Notable for their negativity were articles published by *Newsday* on Long Island, which was curious given that New York had no pending projects. These were lengthy attacks on the Corporation's management, seemingly drawn exclusively from the Synar hearings on the Santa Rosa Project (see the earlier discussion), but only presenting Synar's views and omitting excerpts from the SFC testimony or the dissenting views of other members of the Committee.

One article is worthy of particular attention given its wide audience, the stir it occasioned, as well as its deliberate mischaracterization of subject matter: it was penned by columnist Jack Anderson and published on August 23, 1983, in *The Washington Post* and elsewhere nationwide.

"Synfuel Hunters Explore Saunas and Nightclubs"

The U.S. Synthetic Fuels Corp. hasn't yet found a practical replacement for fossil fuels, but it's not for lack of looking. Synfuels executives have been diligently exploring golf courses, sauna baths and nightclubs around the world.

As has been reported earlier, the Synfuels brass are exceedingly well paid for their unproductive efforts; some of them earn more than cabinet secretaries. Their offices in downtown Washington are elegantly furnished. And when they junket all over the map they take their taste for life's luxuries with them.

The publicly funded corporation's travel expenses amounted to almost $600,000 for 1981-82. My associates, John Dillon and Corky Johnson, combed through hundreds of pages of Synfuels travel records. Here are just a few examples of the corporation executives' sybaritic extravagance at the expense of the American taxpayers:

Four members of the board and five corporation executives took a two- week trip to South Africa last year to visit a synthetic fuel plant. The bill for former Synfuels president Victor Schroeder alone came to $4,290. Both he and board Chairman Edward E. Noble flew first class to and from South Africa. Before the junketers left, the corporation's inspector general wrote a memo criticizing the unseemly size of the South Africa party, but the criticism was ignored, until last week. [Schroeder resigned last week in the face of criticism about agency expenditures.]

Leonard Axelrod, vice president for technology and engineering, is clearly the Marco Polo of Synfuels. Some months he is away from his office for more than 10 work days. He flies to energy industry meetings, conferences and synthetic fuel sites—and a surprising number of the get-to-gethers are held in posh resorts. In April, 1982, for example, Axelrod spent four days in a $160-a-day room at the Americana Canyon Hotel in Palm Springs, Calif. He played two rounds of golf and attended a National Council of Synthetic Fuels Production meeting. In August, 1982, he spent two days at the Tamarron resort in Durango, Colo., at $103.95 a day. While there, he played golf and attended a Midwest Gas Association conference.

Schroeder and his wife, Kathryn, a Synfuels employee, spent eight days in Japan last fall. Among the items in their expense files were bills for a massage, a health spa, camellia plants for their room and the use of a hotel "mini bar." Two of the eight days were set aside for sightseeing. The purpose of the trip was to confer with Japanese businessmen and energy officials. The tab for the Synfuels then-president's trip was $9,082. A spokesman said Mrs. Schroeder took vacation time for the trip and paid her own way.

In October of both 1981 and 1982, the peripatetic Axelrod took two-week trips to London, Brussels and Düsseldorf to attend annual symposiums. His wife, Karen, accompanied him on the 1981 trip, but Axelrod said he paid her fare and lodging. He pointed out that taking his wife along saved the corporation money, because he got a better deal on a double room.

In May, 1982, Axelrod and his wife took the train to New York and back, billing Synfuels $272 for the tickets. An alert staffer wrote a note with the expense voucher regarding Mrs. Axelrod's ticket: "Shouldn't Mr. A. be paying for this one?"

One notation on an Axelrod expense form listed "entertainment" at the Four Seasons Lounge in Houston and described the reason for the expenditure as "technical discussion." The lounge does dispense alcohol, of course, though generally not of high enough octane to be used for fuel.

The reader can judge the merits of some of these points regarding salaries, quarters and travel in light of background material presented earlier in the chapter and by a closer examination of specific examples cited by Anderson to backup his charge of "sybaritic extravagance at the expense of the American taxpayers." SFC staff briefed Anderson's colleagues regarding the facts underlying the article before its publication. Points of rebuttal presented in the following paragraphs are taken from a letter sent by William Rhatican to Jack Anderson.[161]

Anderson had characterized a trip taken by four members of the Board and five senior staff to visit the SASOL facilities in South Africa as a junket. This trip took place just after a new Board was confirmed and that Board was trying to set its own imprint on solicitations that defined the type of projects being sought, the qualifications desired of sponsors, and most importantly the construction and startup experience of the most comparable synthetic fuels facility anywhere on the planet. The SASOL, by far the largest synthetic fuels plant in the world, was producing at levels nearly sufficient to meet half of South Africa's liquid transportation fuels requirements. While those facilities were strongly patterned after the German plants of the Second World War, the scale up of the technology had presented the South Africans with substantial difficulty. They were not able to reach design production levels for about three years. Would the Corporation and its Board not have been greatly remiss in not trying to learn from the SASOL experience?

In citing Schroeder's resignation, Anderson implied that it was in the face of criticism about agency expenditures. Referring to the earlier discussion regarding Schroeder's management, one can see that criticism of Agency expenditures was not an issue. Indeed, during the Corporation's lifetime, it did not use more than 21 percent of the funding that Congress had made available for its operations, and during Schroeder's tenure as President, the percentage was less.

He criticizes Len Axelrod's heavy travel schedule whereas he should have commended Axelrod's diligence. Axelrod, as vice president for T&E, made it a point to be deeply familiar with any project that was likely to receive assistance. In addition, he met with major industry groups as part of the SFC's overall effort to expand the base of firms willing to invest large sums in synthetic fuels and to demonstrate the Corporation's seriousness during its startup period when it was attempting to gain the industry's confidence. The vast majority of his trips were not to the scenic spots described in the article. And, of course, where the locales were pleasant, these were chosen by the industry groups and not by Axelrod. Should he not have played golf? He paid for the golf fees himself. With regard to the Amtrak trip to New York, he only requested reimbursement for his own ticket.

Schroeder traveled to Japan to ascertain if the Japanese, who were interested in synthetic fuels inasmuch as they import virtually all of their oil, would be interested in investing in synthetic fuels projects. Given the difficulty that the SFC was experiencing in finding qualified sponsors willing to invest substantial sums in synthetic fuels plants, Schroeder, as president, would have been lax not to explore these opportunities. The Japanese did invest in the Cool Water Project, albeit not as a result of Schroeder's efforts.

Be that as it may, the specifics cited for Schroeder's extravagance were misrepresented (despite the facts being properly presented to the columnists aides by SFC staff). Schroeder paid for the massage out of personal funds, and he did not have camellia plants and a mini-bar in his room (these were merely names of the hotel's restaurant and coffee shop, whose meal charges appeared on the bill). The final contentious item was entertainment in the Four Seasons Lounge, which presumably gave rise to the article's title referring to nightclubs. The tab for this sybaritic living amounted to $16.58. Axelrod was meeting with senior technical representatives of firms, and to avoid any appearance of conflict, chose to pay for a round of drinks, rather than accept these from industry representatives, who would hopefully be doing business with the Corporation.

All in all, the article was an un-newsworthy tissue of misrepresentation.

Overall, the U.S. media did the Corporation and the public a disservice either by failing to provide factual coverage of a nationally important endeavor. One consequence was that the media has left a historic "black hole." If one were to try to write a history of the Corporation today, two decades later, relying on the archives of the print media, it could not be done: there are only urban myths and disconnected facts to be found. During the summer of 2008, when energy once again came to the forefront of the public's attention, a number of references to the Corporation were made in *The Wall Street Journal*, *The Washington Times*, *The Washington Post*, and the *National Review* (see Chapter 14). For serious publications, they were astonishingly ill-informed with regard to the projects funded, how they performed, what was learned environmentally, and how much was spent by the federal government. The writers did not approach accuracy, which is unsurprising inasmuch as they had nowhere to turn for the truth.

So, while the SFC was slowly but surely accomplishing its new circumscribed mission, its opponents were systematically over many years undermining its public standing. To be sure, the Marshall-Plan ambitions of the Energy Security Act were unrealistic and the Corporation's initial progress was unexpectedly halting as a result of the periods in which it lacked a board with an operational quorum. These factors opened the door for an implacable environmental movement viscerally opposed to expansion of the production of fossil fuels to draw on allies in Congress and the media to undo the Corporation. This is how Washington operates and in itself was unsurprising. Nonetheless, the systematic disregard of facts and the cavalier treatment of the reputations of the SFC's leaders were dismaying—even for the nation's capital.

In any event, the political toll taken by attacks on the SFC became suddenly evident when Congress voted a premature termination of the Corporation—described in the next chapter.

XIII

The Denouement

Chapters 10 and 11 described the Corporation's many accomplishments. It was close to providing the nation with a diverse array of technological alternatives for accessing its vast solid energy resources to provide alternative fuels when they should be needed at a modest cost. Yet as will be seen in this chapter, Congress completely reversed course in just a year's time to prematurely terminate the Corporation. The prior chapter laid the groundwork for understanding this otherwise politically inexplicable turn of events. The ongoing pressure from dissident elements in Congress combined with an unenthusiastic Administration as well as the loss of interest in energy matters countrywide seriously sapped political support for the SFC. This chapter will show how dissident elements gained the upper hand over supportive congressional leadership to undo the Corporation and to cripple the development of energy alternatives.

It will examine: how the program's denouement was foreshadowed in the debates of 1984 that curtailed the Corporation's mission and funding, the nature of the final debates in 1985, and the legislation that terminated the SFC as it transferred the administration of the program of already ratified contracts to the Department of the Treasury.

POLITICAL EVENTS OF 1984

As described in Chapter 10, the SFC found itself in April 1984 without a quorum of the Board to transact business. Normally, that situation would not have been of undue significance in that, under law, the president was obliged to nominate new Board members. So, it should have just been a matter of time until the quorum was restored. That sanguine reading, however, would not have taken account of David Stockman, the director of the Office of Management and Budget. He notified Congress that the president would not submit new nominees until Congress had reduced the obligation authority available to the Corporation from the then existing $16 billion or so by $9 billion.

Stockman's move opened the door to the opponents of the Corporation to go beyond reducing funding to the SFC and attempt to terminate it entirely. Such efforts are evident in remarks by two of the dissidents. Howard Wolpe (D-MI) noted,

> To those of you who have suggested we should let the normal legislative process run its course, I would point out that last year I introduced legislation that 230 Members of this body cosponsored that would prohibit the Synfuels Corporation from making any further allocations until the Congress had an opportunity to review and approve the spending plan. Not only did a majority of the Members of this body cosponsor that legislation, but also a majority of the Members of all three committees to which that bill was referred lent their names as cosponsors. Yet that legislation simply has not moved. This floor vote is the only chance we will have to vote against the wasteful spending of the Synthetic Fuels Corporation.[162]

During the same debate, Congressman Thomas J. Tauke (R-IA) stated,

> [F]or the last year and one-half there have been members of our committee, including the ranking member, the gentleman from North Carolina (Mr. Broyhill) who have tried to get the Synthetic Fuels Corporation dissolved. But what has happened? Has there been a subcommittee markup on the bill? No. Has there been a subcommittee hearing on the legislation? No. It is interesting that the chairman of the committee [citing Congressman Dingell's earlier remarks in the

debate to be presented later in this chapter] comes here and tells us that we ought to get rid of the Synthetic Fuels Corporation when for the last year and one-half a majority of the members of the committee have wanted to do that and we cannot get any action. The bottom line is that when we get down to talking about taking action on this issue the leadership on the other side of the aisle circles the wagons to protect this Corporate welfare.[163]

It was certainly a valid point that congressional leadership on both sides of the aisle was protecting the Corporation. Veteran observers of Capitol Hill know well that the House Majority leadership rules with an iron hand. Thus, subsequent events were especially alarming to supporters of the SFC: the opponents of the SFC pulled off a rare parliamentary maneuver circumventing the desires of the leadership. This occurred in a House debate on July 25, 1984, leading to the defeat of a rule from the Rules Committee regarding the Department of the Interior and Related Agencies Appropriations Bill.[164] In effect, that vote allowed amendments to be proposed in a floor debate that could reduce the amount of funds that had previously been appropriated to the SFC.

Given the rarity of such actions, the accompanying debate provides a good measure of the sentiment of the House regarding the Corporation. Also, this debate can be viewed as the initial attack that reached its culmination a year later.

The Key Players

Who were the major players in this battle? Congressmen Dingell and Synar stand head and shoulders above others who opposed the Corporation. After all, as chairmen of their respective committees, they had the greatest leverage and visibility. In the House, the staunchest champion of the SFC was Jim Wright (D-TX), who was then majority leader. He never wavered and he always made sure that reasoned arguments regarding national energy security stayed at the forefront of the debate. Some committee chairmen also played important positive roles, notably Representative Fuqua (D-FL), Chairman of the House Committee on Science and Technol-

ogy. There were, of course, individual congressmen very supportive of the Corporation, not surprisingly from coal and shale states,
such as Congressman McDade (R-PA).

In the Senate, Senator McClure (R-ID) played the key leadership role in favor of the Corporation among a group of supportive
Senators including Senators Byrd (D-WV), Domenici (R-NM), Ford
(D-KY), Bennett Johnston (D-LA), and Armstrong (R-CO) — representing both sides of the aisle. The Senators opposing the SFC who
frequently arose in the debates included Senators Metzenbaum (D-
OH) and Bradley (D-NJ).

The Essence of the Debate

The arguments raised in debate tended to be the same, year in and
year out. Supporters of the Corporation cited the need for energy security; opponents cited mentioned management shortfalls,
wastefulness and the budget deficit:

Those arguing in favor of the SFC focused on:

The desirability of having proven alternative sources of energy
for enhanced national energy security.

The long lead times required to develop synthetic fuels technologies.

The continued vulnerability of the United States to interruption
of large import levels.

The detrimental effects of undependable governmental policy
that encourages heavy investments by the private sector, only to
reverse course just a few years later.

Those arguing against the SFC emphasized:

Poor management of the SFC.

High salaries and lavish offices granted to SFC employees.

The slow pace of the SFC awarding contracts.

The unreadiness of synthetic fuels technologies for commercial
development/ the need for further R&D.

The budget deficit.

The fact that the synthetic fuels program alone would not provide national energy security.

Representative arguments employed by the Corporation's opponents are well captured by those made by Congressmen Dingell, Synar and Conte (R-MA). Consider remarks by Congressman Dingell:[165]

> Mr. Dingell: Mr. Speaker, I rise, as I rarely do, in support of a rule[166].
>
> Now, my colleagues will remember that some years ago when the Synthetic Fuels Corporation legislation was before this body, I was about as unpopular as an illegitimate child at a family reunion. That was because I felt that the bill at that time went too far and that it unwisely afforded access to the public treasury for enormous sums of money to people who would not spend it correctly. Those expectations have been realized, but still I urge my colleagues to support this rule—for a very special reason. I urge them to support it because when you are out to do in something as irresponsible and as evil as the Synthetic Fuels Corporation, you should do it thoroughly, and you should do it all the way. You should not do it by halves. You should not do it by thirds.
>
> Now, my colleagues here know that I have spent considerable time in the Subcommittee on Investigations and Oversight of the Energy and Commerce Committee investigating the Synthetic Fuels Corporation and the way that its Board of Directors and officers have behaved. Their behavior has been both shameful and shameless. They have been exceedingly careless with the public money, but they have been extraordinarily careful to live well at the public expense, while contributing nothing to the public wealth.
>
> Now, these rescissions which would be offered by my good friends and colleagues, and I commend them for their concern, would cut the amount of money that can be spent by the Synthetic Fuels Corporation back to $2 billion, or to some slightly larger sum of money. That is probably a desirable thing, but those rescissions will still leave in place the Synthetic Fuels Corporation.
>
> Now, the Board of Directors and the employees of the

Synthetic Fuels Corporation remind me of a very highly pampered congregation of well fed, well cared for hogs. They have slopped exquisitely well at the public trough and they needed apparently very little in the way of instruction on how it is that they should live well at the public expense.

Let me share with my colleagues just a few of the excesses we have uncovered at the Synthetic Fuels Corporation.

Salaries. In an unrestrained orgy of spending, Synfuels officials have lavished monies on themselves in high salaries, outrageously generous fringe benefits and luxurious headquarters. Eight Synfuels officials earn more than Cabinet secretaries. Five others are paid at the $69,000 Cabinet level, and 55 of the agency's 177 employees make more than $50,000 a year. The Synfuels Corporation president earns $135,000 a year, and one vice president makes $108,000.

Fringe benefits. All Synfuels employees are allowed to sock away 6 percent of their salaries in a savings-retirement plan, to which the Federal Government contributes 50 percent more. The SFC also pays the full cost of medical and dental insurance, and vests employees 50 percent with retirement benefits within 6 months of their employment - and fully vests them within a year. These benefits are unheard of in other Government agencies or for that matter in the private sector.

Luxury accommodations. The SFC is headquartered in four floors of prime office space in downtown Washington. The building is equipped with saunas, as well as squash and racquetball courts. Synfuels [entered into a] five-year $10 million lease! To achieve the appropriate degree of splendor, $522,919 was spent by the SFC to refurbish their headquarters. The costs included $14,661 for the services of an interior decorator, $374,739 for furniture, and $83,260 for carpeting the executive suites.

Travel. Synfuels executives have been diligently exploring golf courses, sauna baths and night clubs around the world. The Corporation's travel expenses amounted to almost $600,000 in 1981-82. The vice president for technology is clearly the Marco Polo of Synfuels. Some months he is away from his office for more than 10 working days. He flies to industry meetings in such posh resort areas as Palm Springs, Aspen, London, Bermuda, Brussels, and Düsseldorf.

If the rule is rejected and if the amendments are accepted, you will have eliminated potentially a few projects, but you will have left this well-bred congregation of fatted and well-living hogs slopping well and comfortably at the public trough.

The facts regarding most of these points were addressed in the prior chapter, but it should be noted here that all of the SFC salaries and expenditures were within the original intent of Congress: SFC salaries averaged less than other comparable federal agencies and the fringe benefits ran approximately one-half of that given the civil service; over its existence, the SFC used only 21 percent of the funds made available by Congress for administrative purposes;[167] it paid rent comparable to average new construction in Washington, and it never refurbished its quarters (the costs cited were the amounts to build out the unfinished space in the first place). Dingle was well aware of these points given the hearings he held, the GAO investigations he launched, and the voluminous data requests he made over a five-year period.

The next representative example is from Congressman Synar from the same debate, presented here because he was chairman of an important subcommittee, and also because he raises a variety of arguments. Because his remarks were lengthy, they are not included in their entirety, omitting parts dealing with the question of waiving the rule and those that were repetitious.

Congressman Synar's remarks were similar:

Mr. Synar: Mr. Speaker, over the last 2 years I have had the responsibility as chairman of the subcommittee on Environment, Energy and Natural Resources of the Committee on Government Operations to have the oversight over the Synthetic Fuels Corporation. I have also served as ranking member of the Subcommittee on Fossil and Synthetic Fuels of the Committee on Energy and Commerce, where we have authorization over this same corporation.

... This issue does not involve energy security. If it does then I think the Arabs are going to find it very interesting that we are prepared to spend $92 a barrel for oil and we are inviting them to raise the price to that level.

This is simply a question of corporate welfare. ...

Very simply what we have here is the failure by the Synthetic Fuels Corporation, plain and simple. We cannot and we will not meet the goals this Congress set for this corporation.

But more importantly, synthetic fuels are not commercially viable now. They are not commercially viable now. If they were, the industry would be embracing this corporation. But in reality they are running from it.

The issue here today is, Mr. Speaker, very simple: what are the priorities of this Congress going to be? This year alone 3 million children will be without a school lunch program, 300,000 families off welfare, 1 million people less receiving food stamps, and yet we are about to put $2.7 billion into one corporation, $1 million a day for six years.

This is not the type of priorities I think this Congress wants. If you are from Texas and Oklahoma, I want you to go home after this vote and I want you to tell your oil producer that you are prepared to pay $92 a barrel for oil when all they can get is $25 a barrel. ...

We are not killing this corporation because all of us who are concerned about it are committed to the synfuel development in this country. But we have made a commitment of $3 billion up to this point and whichever amendment is allowed to come on this rule will still allow $3 billion more. If $6 billion is not a commitment to synthetic fuels, I do not know what is. ...

...I will stand second to none in concern for this Nation's energy security. I have devoted a lion's share of my time here in Congress to energy issues—in particular the security of this Nation's energy supplies. The issue before us today, however, is not an issue of energy security. It is an issue of how that energy security is most productively achieved at the lowest possible cost to the taxpayer. The

question is not whether we need an insurance policy. The
question is what kind of coverage that policy provides and
how much that policy costs. In my view the Synthetic Fuels
Corporation represents little more than marginal security
- at best - but at a staggering and mostly unnecessary cost.

   ...I believe that I have followed the Corporation's tri-
als and tribulations as closely as any other Member of Congress. There can be little argument that the SFC's 4-year history has been fraught with incompetence, mismanagement
and cronyism. And throughout this period I have often been
a vocal critic of the program's direction, or lack thereof. ...

Synar seemed not to have had a real grasp of the issues as seen
in his corporate welfare argument, which is at the heart of his position. As noted in the previous chapter, the $92 per barrel figure
cited for the Unocal project was taken out of context, being a hypothetical number assuming a number of years of inflation that might
or might not occur. The contractual price guarantee figure was between $60 a barrel and $67 a barrel. If one were to calculate the
actual subsidy to Unocal per barrel to be produced from the facility
in present day dollars (as the Corporation did), the subsidy would
have amounted to between one and two dollars a barrel.[168] This was
carefully explained to Synar's staff before this debate.

His comments about his commitment to synthetic fuels in general and against the Corporation's mismanagement are belied by
his performance as Chairman of his committee over a five-year period.

One more example of the reasoning of opponents is taken from
the remarks of a Republican congressman, Silvio Conte of Massachusetts, who was actively pushing for the waiver of the rule under debate (the portions of his remarks dealing with the procedural
question are omitted here):

Mr. Conte: ... I had proposed and printed in the RECORD,
an amendment providing for a $9 billion synfuels rescis-
sion; Congressmen Wolpe and Synar had proposed a $10.2
billion rescission. Whatever the amount, the important
thing is that the synthetic fuels program has been like an

unguided missile, with too much money landing on too many ill-conceived projects. We need to defeat this rule so that an amendment can be offered to cut back on this disgraceful waste of taxpayers' money.

Mr. Speaker, synthetic fuels were the wave of the future back in 1980. But since then, oil scarcity has turned to glut and the price of imported crude has dropped more than 25 percent. Oil imports are down 33 percent, and the Strategic Petroleum Reserve contains over 410 million barrels of oil. Circumstances have changed, but the Synthetic Fuels Corporation has not.

To my friends who want to see a continued synthetic fuels program, I say that … the only way to save the corporation is to cut it back.

Just look at the list of projects funded by the Corporation and you see a consistent pattern of waste and abuse. [*At this time, of course, the SFC had funded only Cool Water and Dow Syngas, both of which turned out to be successes.*] You see sweetheart deals, outdated technologies, shaky private support, and obscene price supports. A GAO report released yesterday revealed that the Union Oil shale project will be getting a million dollars a day in price supports for 6 years. And that's not all! It's also getting … billion in tax breaks!

That's the kind of obscenity you expect from *Penthouse*, but it can't be tolerated from a Federal corporation.

…My colleagues over on this side of the aisle are going to campaign this year on the question of fairness, on the question of corporate welfare, on the question of balancing the budget, and on the question of environmental protection. Well, as I said this morning, now is the time to put up or shut up!

A final example from the debates of 1984 comes from Senator Metzenbaum's remarks made on September 26, 1984, regarding appropriations language that would cut back the funding of the Corporation.[169] This adds a few additional arguments to those already summarized, and is especially illustrative of the cumulative impact of the attacks made on the Corporation over the years:

Mr. Metzenbaum: The fact is that the history of the SFC is of an agency crippled by conflicts of interest and mismanagement.

Two presidents resigned amid charges of conflict of interest and mismanagement; five Board members have resigned; seven top corporation officers have resigned; awards were made to projects after a Board member with a conflict delayed his resignation to approve them.

When Congress created the SFC in 1980, it went overboard to insulate the agency from public scrutiny. The scandals and outrages that have subsequently bedeviled the SFC are the direct result of this lack of oversight and accountability.

If we are to avoid the shenanigans of the last three years, if we are to finally develop sound synfuels projects, we must make the SFC more accountable to the American people... .

Here the reader can see the payoff from past political attacks against the leadership of the Corporation. Chapter 12 discussed the circumstances regarding the resignations of Schroeder and Thompson from the presidency; nowhere was any scandal evident. The other Board members who resigned did so because their terms were limited and expired (two) or because they received other presidential appointments. The officers who resigned did so because the Corporation was always viewed as a short-lived institution with which one did not expect to make a career. In four years, some turnover was inevitable.

Metzenbaum's charges of lack of SFC oversight seem oblivious of extensive Congressional hearings, GAO investigations, and Congressional requests for information routinely undertaken. Moreover, the SFC was just as open to the scrutiny of the American people as any government agency. The SFC had to follow the basic requirements of the Freedom of Information Act and had no exemptions beyond protecting proprietary data. Similarly, the Energy Security Act required Board meetings to be open to the public—much as under the Sunshine Act—except when proprietary information or information sensitive to the competitive selection process was discussed. Finally, Vic Thompson's resignation was not delayed to vote on a project on which he had a conflict. The reader may recall that his purported conflict was indirectly associated with the Chapparosa Project. The vote in question had to do with the Dow Syngas Project.

Illustrative Arguments by Proponents

While there were many articulate supporters of the Corporation in the House and the Senate, just two of their remarks are cited at this juncture, largely because of the importance of the speakers and the completeness of their arguments. The first section quoted is Majority Leader Wright's remarks on July 25, 1984, regarding the waiver of the rule and Senator James McClure's (R-ID) remarks in the September 26 debate (he was Chairman of the Senate Energy and Natural Resources Committee):

> Mr. Wright: Mr. Speaker. I rise in support of the rule. I think it would be a gross mistake for us to signal, only 5 years following the Arab oil embargo and the long gas lines of 1979, that we have forgotten all about that and that we are now in retreat from our bold commitment to energy independence.
>
> Just think what signal that would send to the Qadhafi's and the Khomeini's of this world—that we no longer are serious about making this Nation of ours invulnerable to their pressures.
>
> Let me show you what we are talking about. Synthetic fuels are the way in which the United States can make itself invulnerable and independent of those nations which wish us less than well. By God's grace we have enough resources in this country, if we have the wit and the will to develop them on our own, that we can be energy independent.
>
> We are going to run out of oil and gas. Experts do not disagree. They may disagree on how long. But within one generation or two at the most, the experts in geology tell us that our voracious appetite for energy and power will have exhausted all those resources that it took nature some millions of years to lay down under the earth in the form of oil and gas deposits.

But what do we have? Out in the Rockies of Colorado on the western slope, we have as much oil in the shale rock as exists in the Persian Gulf. Think about it. These four vials that I hold in my hand offer a clear demonstration of what has been developed under the Synthetic Fuels program.

Here in the first vial we see rock cracked from the shale on the western slope of the Rockies. In the second vial, that cracked rock has been retorted; it has been subjected to a high pressure, high intensity heat and reduced to an ash.

The third vial shows the crude oil which that crushed ash yields. In the final vial, we have the golden distillate from the refined crude oil.

There is enough of this in the Colorado Rockies to replace all the oil in the Persian Gulf. Are we foolish enough not to develop efficient means of extracting it before we run out of oil and gas reserves?

Let us think about coal. The United States could be the Middle East of the world in coal. We have about one-third of all the world's known coal reserves. This little lump of coal which I hold in this second display contains every property of crude oil. This vial contains crude oil gained from crushing and exposing coal to a high-pressure process; hydro coal. The liquid in that little vial is a high-grade, low-sulfur content oil. This was produced 12 years ago. It is not that we do not have the technology; it is not that we need research. We need to develop it. Is it too early? There are those who say, "We do not have the time to develop a commercial product." Well, the Germans developed a commercial product as long ago as World War II. They kept the Luftwaffe and the Wehrmacht operating for many, many months after they would have been out of business had it not been for their very crude processes of using that soft, brown coal from the Ruhr Valley, and making high octane aviation gasoline out of it. It is not anything new. We were warned as early as 1952 by the Paley Commission that we were going to run out of oil and gas and if we had any vision, we would have begun then developing commercially usable synthetic fuels.

We did not do it but South Africa did. With only one forty-fourth of our gross national product, that little country already has developed commercially viable synthetic fuels from coal. They make gas, butane, propane, plastics, fertilizer. They make everything that we can make out of crude oil. They do not have any crude oil but they are not vulnerable to the Arabs. Have we less vision than they? I agree that those who have been appointed to the Synthetic Fuels Corporation have not behaved in the way that they should. Their sin has not been that they spent too much; but that they have done too little. They were commissioned by this Congress with a mandate to develop actual production of not less than 500,000 barrels a day by 1985 [*sic:* 1987] and not less than 2 million barrels a day by 1990 [*sic:* 1992]. Their sin is that they have not prosecuted the program vigorously enough. Their sin is that they have ignored the congressional mandate. Someone said, oh, we have let $6 billion to go forward in loan guarantees.

Now, to rescind loan guarantee money is not to reduce the deficit by one thin dime, any more than if we were to rescind the money for FHA loan guarantees. You would not save any money; you would build fewer houses. That would be the result of this. We would not save money, we would stretch the time getting to our goal of energy independence. If you think that $6 billion available in loan guarantees is too much, stop to think that we are being drained by $50 billion every year because of our vulnerability to Arab oil. In 1941, we had vision. We ran out of rubber. The Japanese took over our rubber supplies. Franklin Roosevelt called Bernard Baruch and Bill Jefferson together and said we must develop synthetic rubber. Nobody knew how to do it; we did it because we gave it enough priority.

We made a crash program out of it, and 3 ½ years later, when the Allies rolled into Berlin, they rolled on rubber tires made from an indigenous American synthetic rubber industry. We can be that nation again. That is the message we need to send to the Khomeini's and the Qadhafi's of the world. Not that we are so penurious that we worry about a little bit of money in loan guarantees and quail at the relatively small cost of becoming energy independent. They are not afraid of our weapons; they might be afraid of our energy independence. And in the long run, our independence as a nation depends as much upon that as it does upon the weapons.

Senator McClure's remarks are taken from both the September 26 and October 3 debates:

Mr. McClure: ... Mr. President, I rise today to speak on the future of the Synthetic Fuels Corporation program. There is a continuing need for completion of the principal objective of statutory phase I, namely, the accelerated development of a selected number of commercial scale plants located in the United States.

Events in the Middle East continue to argue for increased efforts toward ensuring energy security for the United States. We have made commendable progress in the area of energy conservation. We have also made considerable progress in filling the Strategic Petroleum Reserve. But the specter of supply disruptions, accompanied by precipitous price increases, and foreign policy blackmail is always on the horizon.

The need still exists to ensure a long-term national capability for the domestic production of synthetic fuels. Such a capability for replication is needed as an insurance policy against an ever-widening gap between domestic consumption and production. And such a capability is needed if our abundant reserves of coal, oil shale, and tar sands are to fill this developing domestic supply gap in an environmentally compatible way.

But there is a complacency in the United States regarding the seriousness of our long-term energy vulnerability. The current ready availability of petroleum supplies at reduced prices is misleading. The suggested abundance of supplies has diverted our attention from the fact that over the longer term these supplies are politically and economically insecure. And the lack of secure energy supplies is, indeed, a grave national security problem.

Our complacency, while unwarranted, is understandable. As a Nation, today we are far better prepared to cope with disruptions in the world supply of petroleum than in the recent past. The Strategic Petroleum Reserve now has over 100 days of imports. We also are importing approximately 28 percent of our oil needs.

As a consequence, there is an ongoing annual trade deficit of $60 billion annually from crude oil alone. In addition, there is the cost to fill the Strategic Petroleum Reserve as well as $40 billion for a Rapid Deployment Force to ensure supplies from the Persian Gulf.

But this situation will change dramatically over the next decade, as it has over the last decade. By the turn of the century, we could be importing more than 50 percent of our petroleum unless we develop alternative sources of domestic supply. And synthetic fuels is one of those alternatives. If we are to reduce our strategic vulnerability we must develop such alternatives.

Recent events in the Persian Gulf only serve to dramatize the vulnerability of our present situation. They accentuate the importance of developing, in advance, a synthetic fuels option for the United States. The eyes of OPEC, the free world, and the Eastern Bloc all remained focused on the SFC and whether we are capable of keeping our national commitment to the objectives of the Energy Security Act. If we falter in our commitment to the objectives of a viable synthetic fuels capability in the United States, we would send the wrong signals to our allies and OPEC.

For several months the Synthetic Fuels Corporation has been handicapped by the lack of a quorum for the conduct of business. Since the SFC was declared operational by President Reagan, a series of solicitations have been issued in order to achieve the objectives of statutory phase I. The proposals that have responded have effectively defined the universe of potential candidates.

The SFC has developed a basis [*sic*] business plan. The

projects already included in the plan will have a production capability by 1989 of 135,800 barrels of oil equivalent a day. But, more importantly, these same projects will be expandable on the same sites to a production capability of almost 400,000 barrels of oil equivalent a day. Even without expansion, the total production from the 11 projects for which the SFC has awarded contracts of [sic] authorized letters of intent is estimated at over 1 billion barrels of oil equivalent over their useful lives.

Each and every one of these private sector initiatives were advanced in good faith in response to a national imperative. Current private sector commitments to the SFC's Basic Business Plan are an estimated $11.4 billion. An additional $3.78 billion is being projected for contingencies and other projects. When all of these non-government-owned projects have proceeded to completion approximately $14.77 billion in private sector funds will have been invested in America's energy future.

When the SFC had an operating quorum, it is reported that the project sponsors were expending approximately $13.4 million per week. In the absence of a working board quorum these private sector monies are now at risk.

Beginning in December 1982 (H. Rept. 97-978), the Congress in the conference report on the Interior Appropriations Act for fiscal year 1983, expressed concern with respect to the slow progress of the SFC in awarding financial assistance. Subsequently in November 1983, during consideration of the Further Continuing Appropriations Act for 1984, the Congress reiterated its concern and reasserted its intent and commitment to the "goals and objectives of the Synthetic Fuels Corporation as expressed in the Energy Security Act."

In addition, it was the judgment of the conferees that, "the Synthetic Fuels Corporation as the appropriate federal entity to assist and facilitate early production of synthetic fuels." That is still the judgment of the Congress.

The private sector approach intended by the Congress in enacting the Energy Security Act has been achieved and maintained by the SFC. Currently there are billions of dollars of private capital on the table. This equity was raised by the private sector in response to the national security imperative that underlies the SFC's program. And this equity may well be lost if the program does not proceed.

In addition to the national security benefits, there are social and economic benefits to be realized from creation of a synthetic fuels infrastructure in the United States.

There is considerable environmental protection experience that is to be gained from the permitting and operation of these initial projects. The SFC's environmental program offers great potential for mitigation [sic] any adverse environmental consequences from deployment of such technologies.

Synthetic fuels are in fact an integral part of any solution to our present or future environmental problems. But sound environmental management can best be achieved under a deliberate program, unhastened by the crisis atmosphere that would occur if there were a major disruption of oil imports.

There are three important environmental reasons to develop synthetic fuels technologies:

First, no proposals will receive assistance from the SFC unless it can demonstrate that it can obtain all required environmental permits as dictated by Federal and State law. If such permits cannot be obtained, the SFC's policy dictates that such projects not receive any assistance from the Corporation.

Second, before a binding commitment for financial assistance can be made there must be established a sound environmental monitoring program, with the assistance of the EPA, the DOE, and State regulatory agencies.

Third, synthetic fuels are an integral part of the solution to our present and future environmental problems. For example, synthetic fuels technologies, such as those employed by the Cool Water project in California, offer the potential for substantially reduced atmospheric emissions of sulfur oxides and nitrogen oxides compared to conventional means for the generation of electricity from coal. This technology also can be readily replicated by many of our Nation's utilities in smaller units in rural as well as urban areas. There also is the engineering and economic experience that is to be obtained from the operation of these initial synthetic fuels projects. And there are the jobs that will be created: First, in the States where the projects will be sited; and, second, in those States where project components will be fabricated. As much as 70 percent of the capital costs of a synthetic fuels plant will be spent on equipment and materials such as steel, cement, pipe, and instrumentation. These requirements will both maintain existing jobs and create new jobs. According to the Bureau of Labor Statistics for every $1 billion spent on heavy construction projects, such as synthetic fuels, approximately 30,000 full-time and part-time jobs are created. Thus the current projects pending before the Synthetic Fuels Corporation would generate over 400,000 jobs.

At this point, he inserted material in the record and outlined the specifics of a proposed legislative amendment. The example continues with further remarks he made on this subject on October 3.

The supporters of the Bradley-Nickels amendment are quick to justify their position by incorrectly characterizing the nature of the SFC price guarantees to the Union Oil Shale project. Let me review with you how much that assistance really is:

First, using the same methods advocated by the Office of Management and Budget, the present cost of the assistance to Union is 98 cents per barrel. When taxes are taken into account the assistance is half this amount.

Second, while the peak price guarantee by the SFC could be $92 per barrel, the SFC would pay the differential of approximately $44 per barrel in 1989. And because the price guarantee payments are fully taxable, the differential actually is a peak of about $17 per barrel today. However, over the life of the contract the assistance as mentioned would only be an average of 98 cents per barrel before taxes, and half that amount after taxes.

Third, as the GAO points out, the Union project will take 11 years to achieve a positive, cumulative cash flow. By that time Union will have invested $1.2 billion of its own money in the project.

Finally, as provided in the SFC contract, there actually will be profit-sharing with the Federal Government. Under this provision of the contract, the Federal Treasury will recoup over $3.15 billion of the assistance provided by the SFC, which is a net cash flow of $450 million.

Mr. President, similar recoupment provisions are addressed as a standard feature of SFC negotiations. As Chairman Noble has noted, if prices rise by only 2.3 percent over inflation during the next 25 years, then virtually all the moneys expended by the Corporation for synthetic fuels development should be returned to the taxpayers. The SFC program, Mr. President, is aimed at providing a diverse spectrum of synthetic fuels production experience on which to base future financial, technical and environmental decisions.

A principal benefit to be gained from this experience will be the operation of such projects in an environmentally acceptable manner. Only full scale, long-term commercial operation can provide reliable experience. But, more importantly, research and development activities alone will never resolve the questions and risks accompanying any first of a kind commercial plants. Yet the Bradley-Nickels amendment would relegate us to continuation of only research and development of synthetic fuels.

The SFC program provides the basis for efficient private sector development and deployment of synthetic fuels in response to market forces. Once the Federal assistance ends the projects must be financially viable. This is the principal feature of the Energy Security Act....

There were many more debates over the next year and a half, but the arguments were only variations on these themes. The reader can make up his or her own mind as to which were the more fact-based.

As summarized in Chapter 10, following these debates Congress passed new legislation cutting the Corporation's funding authority by about half and redirecting its mission to proving a diverse mix of technologies to exploit the nation's largest energy resources, rather than supporting high levels of synthetic fuel production.

POLITICAL EVENTS OF 1985

It seemed during the first half of 1985 that the debate was settled and that, with its newly appointed Board, the SFC could go about completing its mission. Even as late as May, the Senate routinely confirmed the new Board members.

But the hardcore opponents of the SFC had not been mollified. Indeed, there was no way of mollifying them short of doing away with the Corporation. Congressman Dingell's remarks summarized earlier made it evident that he was calling for the complete abolition of the SFC, not just for reducing its funding. And, he had publicly vowed to be the "nightmare" of the Corporation.

The storm suddenly reappeared in July, when the House once again voted to overrule the Rules Committee to debate the appropriations for the Corporation,[170]. While the arguments had not changed from the prior year, all the political guns were brought to bear: Wright spoke at length. So did Majority Whip Tom Foley (D-WA) who was seen as representing the Speaker and who supported the SFC, as did Fuqua and other committee chairmen. The leadership, however, did not prevail. Thus, the door was open for the House to reconsider appropriations for the Corporation, which it did a few days later. The House voted on July 31, 1985, 312 to 111 to rescind all but $500 million of previous appropriations for the SFC, and that could only be used for administrative expenses[171].

The future of the SFC was now in the hands of the Senate. And the Senate sentiment was not as securely attached to proponents as it had been in past years. The outcome would depend on the strategy of the SFC supporters and the posture of the White House. If the Administration should shift from one of unenthusiastic tolerance

of a modest program to one of opposition, it would be all over. Was
Secretary of Energy Herrington's opposition to a Great Plains deal
just an indication of misplaced ideology, or did it signal a position
of the administration?

The leadership of the SFC supporters was once again in the
hands of Senator McClure, the Chairman of the Energy and Natural Resources Committee. Otherwise, the supporters were without
the help of Majority Leader Robert Dole (R-KS), who did not take
part in the debates, but voted against the Corporation in all the key
votes.

One of McClure's first moves was to find out where President
Reagan stood, and so he wrote the White House on October 15,
1985:

The President
The White House
Washington, D.C.

Dear Mr. President:
I know we both are aware of the importance of the program of the U.S. Synthetic Fuels Corporation to our country's energy and economic future. Your appointment of the
Board of Directors under Chairman Ed Noble in 1981 recognized the national security benefits of continuation of this
vital, government-industry partnership.

Last year, however, in recognition of the need for budgetary constraint, we negotiated an agreement for the rescission of $7.375 billion of the SFC's obligational author-
ity and eliminated the synthetic fuels statutory production
goals in the Energy Security Act. As part of that agreement
the Senate confirmed your three nominees, so that the SFC
had a working quorum. Considerable effort was expended
on my part to achieve that agreement, in recognition of our
mutual concerns for the budget deficit.

Subsequently, and again as part of our agreement, the
SFC Board reformulated its business plan and its renegoti-
ating the size of individual projects so as to achieve further
budgetary savings of between $1.2 and $2.5 billion. Equally

important, private industry has invested almost $1.4 billion in good faith to implement last year's agreement.

When completed, this reduced program will significantly contribute to the statutory objective of establishment of the infrastructure necessary to support a commercial synthetic fuels industry in the United States. Moreover, this critical objective will be achieved at less than 10 percent of the cost contemplated by the Congress in the Energy Security Act.

As you are aware, I reached that agreement with Edwin Meese, then Counselor to the President; David A. Stockman, then Director of the Office of Management and Budget; and Donald Paul Hodel, then Secretary of Energy, who were acting on your behalf. Acting on the belief that they were indeed carrying out your instructions, I made representations to a great many of my colleagues that our agreement would be the last effort to alter our synthetic fuels program. Now it appears that the rules have changed since you recently reorganized your staff so that new individuals have assumed these responsibilities. Members of your Administration are actively working to overturn our agreement. To my knowledge you continue to support the SFC's reformulated and restructured program as we agreed. Am I correct in that assumption?

The importance of the SFC's program to our country's energy future was recently supported unanimously by some 34 States who are members of the Western Governors' Association and the Southwest Regional Energy Council. Now, when we are about to culminate a four-year effort to achieve the principal goal of the Energy Security Act—commercialization of synthetic fuels in the United States—is not the time to scrap this critical national program.

Sincerely,

James A. McClure, Chairman

In a floor debate on October 31, Senator Johnston (D-LA) stated his views with more emotion, if less finesse. He said

> We got a commitment by the Administration to stick to that compromise. Now, it is at any given moment difficult to say where the Administration stands on this matter. I know the Secretary of Energy has written a disparaging letter about the Corporation. I do not know where the President stands. I do know they made a deal. They put the word of the administration on the line behind this compromise. ...
>
> Mr. President, it is not just a question of whether those of us in the Senate can count on the administration. There is a vast array of people involved in the synthetic fuels business who have also got a right to rely on what the administration says. ... Mr. President, I think it would be an outrage for the administration to say: "Oh, we were just kidding. We made a deal and gave our word and unfortunately, private industry, your $1.2 billion will now go down the drain because, you know, caveat emptor, you don't have the right to rely upon what this administration says.

A month or so later, the administration sent an ambiguous reply signed by James Miller, III, Director of OMB, indicating that the administration was willing to abide by the expressed will of the Congress. (As we will see later, what the letter meant was that the White House expected the Congress to terminate the SFC without the administration having to alienate the Corporation's supporters.)

The forces for and against the Corporation quickly confronted one another again on the floor of the Senate. The debate erupted on October 31, 1985, with the introduction of two amendments to Interior Appropriations language.[172] Working with McClure, Senator Johnston introduced an amendment designed to forestall worse action that deleted another $500 million from the Corporation's authority and set a deadline of September 30, 1986, by which time the Board would have to conclude making any awards it was going to make.

Senator Metzenbaum, the old foe of the Corporation was ready and waiting. He introduced an amendment that had the same language that passed the House—i.e., to terminate the SFC. Up until this point, Metzenbaum had hidden behind statements that he was for the program but just had not liked the operation of the Corporation. For example, on September 26, 1984, he had stated, "Mr. President. I support a strong, well-funded SFC—but a strong, well-funded SFC is still impotent unless the integrity of its decisions are unassailable. ... [I]f we are to finally develop sound synfuels projects, we must make the SFC more accountable to the American people." He then introduced the requirements for openness of data and meetings, which had been adopted in 1984's legislation.

But in the 1985 debates, he stated, "I originally supported enthusiastically the Synthetic Fuels Corporation. ... But, in this instance, I believe there is no room for compromise and only a total routing, a total retreat, of the Synthetic Fuels Corporation and the utilization of those dollars for more necessary purposes in this country is appropriate."

In the debate, he was strongly joined by Senators Bradley and Evans. No new arguments against the SFC were introduced, but the emphasis this year was on the exigencies of the budget deficit, and on the need for more R&D before building commercial facilities. Both were spurious arguments. As the proponents of the SFC pointed out, because of the structure of the loan and price guarantees, there would be virtually no outlays for another five years, and thus no impact on the budget decision being considered. Earlier chapters discussed why the R&D argument did not apply. (Moreover, in the ensuing years, Congress authorized no new R&D.)

The support of the SFC was principally argued by Senators McClure, Ford, Johnston, and Byrd (the Minority Leader). Given how recent the last congressional action was, the most cogent point was voiced by Senator Johnston, "Mr. President, the past mistakes of the Corporation have, I believe, been corrected. Indeed, we just confirmed three additional members, outstanding in record and experience, who, with the now seasoned survivors of the Synthetic Fuels Corporation, I believe, can, will, and are doing an excellent job." Curiously, none of the opponents took issue with how well the SFC was, or wasn't, complying with the Congressional action of the prior year.

Rather than citing from the rest of the debate, these sections will simply quote from a "Dear Colleague" letter issued by the Committee on Energy and Natural Resources on October 4, 1985.[173] Much of it was similar to the letter sent by McClure to the President and won't be repeated.

Dear Colleague:

In the near future it is expected you will once again be asked to vote on an amendment to abolish the Synthetic Fuels Corporation (SFC), which we urge you to oppose. ...

In requesting your opposition to this amendment we urge that you consider the following recent reports that underscore the United States' continuing economic and international oil supply vulnerability:

The Iran-Iraqi War poses a new and greater threat to U.S. interests, according to the Reagan Administration oil analysts and former Ambassador Richard Helms (*Washington Post*, September 27, 1985);

Iran's oil exports have declined by two-thirds as a result of Iraqi attacks (*Wall Street Journal*, September 27);

Significant oil price increases recently have occurred because of such factors as Iran's and the Soviet Union's temporary suspension of oil exports, and the Saudi announcement that it will take on no new contracts (*New York Times*, September 27); and

Recently, the U.S. Geological Survey concluded that the Middle East will increasingly monopolize dwindling world petroleum supplies and the U.S. can expect to return to the energy crisis of the 1970's, experiencing regular supply interruptions with but a few decades left to enjoy the convenience of crude oil as our major energy fuel and therefore it is essential for the Nation to develop its own alternative energy resources (*Washington Post*, September, 26)....

Sincerely,

J. Bennett Johnston, Ranking Minority Member, James A. McClure, Chairman, Wendell H. Ford, Quentin N. Burdick, Mark Andrews, Gary Hart, Arlen Specter, Alan J.

Dixon, Paul Simon, William L. Armstrong, John H. Warner, Orrin G. Hatch, Jay D. Rockefeller, John Heinz, Ted Stevens, Paul S, Trible, Jr., Pete V. Domenici, Russell B. Long, Jake Garn, Robert C. Byrd, and James Abnor."

With considerable parliamentary maneuvering, Senator Mc-Clure almost pulled it off. His first move was a motion to table Metzenbaum's amendment. While his move failed by a vote of 58 to 41, he managed to get the vote on the amendment itself postponed until December 6, 1985 (five weeks later). At that time, there was further floor debate. When the time appeared right, he called for a vote, and the amendment was defeated by 43 to 40. But it seemed clear that if the 17 members not voting had participated, the Metzenbaum amendment might well have passed.

Thus, when further efforts by McClure appeared to be heading to a Conference Committee with the House in which a compromise would be reached along the lines of the Johnston amendment, the administration acted again and showed its true colors. James C. Miller III sent another letter to the Hill, dated 12 December 1985. This one was addressed to Senator Mark Hatfield (R-OR).[174] It read:

> Dear Mark:
> The Administration has been requested to state its position on whether or not funding for the Synthetic Fuels Corporation should be continued.
> When the Synthetic Fuels Corporation was created in 1980, oil prices were increasing rapidly and were projected to reach $75 to $125 per barrel by 1990. Experts were also predicting future oil shortages and the possibility of increased OPEC control over our energy future. Since oil prices peaked in 1981, the world energy outlook has improved substantially. Oil prices have declined by over 30 percent and because of world-wide overproduction oil prices may go down even further. In addition, our oil imports have declined by over 25 percent since 1980, and the Strategic Petroleum Reserve, now containing nearly 490 million barrels, provides over 300 days of protection if OPEC halted supplies, as opposed to 17 days in 1980.

The Administration believes these fundamental changes make it impossible for a commercial synthetic fuels industry to develop without enormous budget outlays that would not be offset by any economic benefits.

In addition, passage of the Gramm-Rudman-Hollings legislation and our need to reduce the Federal deficit requires us to reduce unnecessary government spending. Accordingly, the Administration no longer believes that continued funding of the Synthetic Fuels Corporation serves any useful purpose. Moreover, the Administration now supports efforts to rescind all funding available to the Corporation and opposes any efforts to use such monies to fund any other programs.

This was hardly prescient. After a hiatus of a decade or so, oil imports increased steadily to all-time highs in the new century in terms of absolute number of barrels per day and of the percentage of consumption. Moreover, the dependence on OPEC and the Middle East was reestablished as well and, even though the Strategic Petroleum Reserve continued to be filled, it could cover only about 80 days of imports. Prices exceeded previous historic highs and the United States had fought two wars in the Middle East. And although Gramm-Rudman called for a zero deficit by 1991, the deficit continued to bedevil Congress.

In any event, with the overt opposition of the President, McClure could not swing the tenuous compromise through the Conference Committee, and the Committee voted to terminate the Corporation.

THE FINAL LEGISLATION

The legislation terminating the Corporation was contained in the Conference Report on House Joint Resolution 465, Further Continuing Appropriations for Fiscal Year 1986. This became enacted in law as Public Law 99-190. The main elements of the legislation were as follows:

The funds remaining in the Energy Security Reserve and not obligated as of the date of enactment were to be rescinded.

The action did not apply to funds made available for clean coal technology programs in the earlier actions, and it made an additional $400,000,000 available for the Clean Coal Technology Program (sums that were not much less than what finishing the SFC program would have entailed and that produced little in the way of new technology).

After this date, the Board could not make any legally binding awards for financial assistance (including any changes in existing commitments), but nothing was to impair the duties, rights and obligations of the Corporation for carrying out the terms of existing legally binding commitments.

Within 60 days of enactment, the Board of Directors were to terminate their duties and be discharged.

Within 120 days of enactment the Corporation was to terminate in accordance with Title J of the Act.

The Secretary of the Treasury was to assume the duties of the Chairman and those responsibilities were not to be transferred to any other Federal department or agency.

After this date, no employee could receive salary in excess of civil service pay and OPM was to monitor benefits for reasonableness.

The Board moved expeditiously to implement the legislation. At this time, there were about 125 employees with the Corporation. An orderly separation schedule was developed. Most employees departed within six weeks. Only those required for shutdown procedures—closing out contracts, personnel separation, boxing of files, and moving project contracts to Treasury— were kept until the termination date. The Board held one more meeting in January authorizing the final actions, and resigned prior to the statutory date. A cadre of senior Treasury employees supervised the actions until April 18, 1986, the last day of the U.S. Synthetic Fuels Corporation.

Eight SFC employees transferred to Treasury to manage the government's residual responsibilities for the four commitments. The final chapter describes the fate of these projects and summarizes the significant experience gained by the nation from their construction and operation.

XIV

Epilogue

So the SFC came to a premature end. While that concludes the saga of the Corporation's founding, existence, and termination, the funded projects continued in their construction, operation, and learning, thereby better arming the nation as it confronted future energy challenges. As noted in the previous chapter, the legislation terminating the Corporation transferred responsibility for the administration of the four financial assistance contracts and the associated project monitoring to the Department of the Treasury. A prior chapter indicated how the Department of Energy assumed the Great Plains Project, which it had originally assisted with a loan guarantee, when that project defaulted.

Let's see how the fates of these projects unfolded, what the net cost to the government was, and what was learned in the areas of technology development, environmental impact, and the use of innovative financial incentives by the government to promote the building of pioneer facilities.

SUBSEQUENT HISTORY OF THE FIVE PROJECTS

When the dust settled, five synthetic fuels projects were built and operated for a varying number of years: Cool Water, Dow Syngas, Great Plains, Parachute Creek, and Forest Hill. Their relative performance varied considerably, but much was learned in each case.

Cool Water
Construction of the Cool Water Coal Gasification Plant began on December 15, 1981, and was completed by May 1, 1984, one month

ahead of schedule. Operation of the coal gasifier began on May 7, 1984, and initial production of electricity occurred about two weeks later on May 20. Commercial operations began on June 24, 1984, upon completion of a stringent 10-day acceptance test run required by the Corporation, and syngas production became eligible for Corporation price differential payments the following day.

Despite encountering various difficulties typically associated with the startup of a first-of-a-kind facility, the plant exceeded its projected production goals by having operated at full design production capacity in terms of coal throughput, syngas generation, and electric power output. From initial startup in June 1984 through the end of 1985, the plant demonstrated an on-stream factor in excess of 55 percent; during 1985, on-stream time exceeded 60 percent; and in the last years of the five-year demonstration period, the plant achieved a capacity factor over 90 percent during many months and its rated capacity was raised to 127 megawatts.

The plant's environmental performance was particularly encouraging. The project obtained all of the environmental permits required to operate the facility. The principal permits were the Prevention of Significant Deterioration (PSD) permit (air pollution), issued by the U.S. Environmental Protection Agency, and the California Energy Commission permit. The latter permit covers environmental and socioeconomic factors associated with project development as prescribed by applicable state and local regulatory agencies with respect to air quality, public health, water quality, solid waste, and worker health and safety. The permits required considerable monitoring of air, water, solid waste, hazardous waste, and socioeconomic impact. Actual monitoring results showed that the project emitted only a fraction of the pollutant levels allowed for conventional power plants equipped with flue-gas desulphurization units (actual data is presented in a later section).

A group of thirty utility companies followed developments at Cool Water closely, while a subgroup of perhaps five to ten utilities undertook studies aimed at the installation of such plants in their systems in later years. The general economics appear competitive with conventional coal fired central station plants having flue gas scrubbing. But, equally important, the technology has advantages for utilities as they confront fundamental uncertainties with regard to changes in the cost of fuel and in the growth of demand for electrical power. The IGCC has a capability for being built on a modular basis—allowing reduced construction time and, thus, a closer match with growth of electricity demand. Moreover, it permits the utility to first install a combined-cycle plant using natural gas that can be followed by the addition of a coal gasifier when justified by rising gas prices.

Despite the sizeable number of participants, the project's organizational arrangements and management structure worked exceedingly well and could be a possible model for future sponsors of projects utilizing new technologies. Specific examples of management achievements include:[175]

The project was completed under budget. At the time the project began the design, the estimated cost for engineering, design, procurement, and the construction of the Cool Water facility was $294 million. This amount did not include the cost of the adjacent air separation plant, which was built by Airco. The as-built cost of the facility was about $278 million.

Construction was completed on schedule.

Only a short period (May 7, 1984, to June 23, 1984) was required for startup of the plant.

Plant capacity has been demonstrated at or above design coal throughput and power plant output.

Longer term operation at above design capacity factor was achieved.

Nonetheless, the sponsors did not proceed with Stage II as had been contemplated under the contract. As mentioned earlier, energy prices weakened throughout Stage I at the same time that the need for additional electricity generating capacity in the Southern California Edison service area was vanishingly small at the time that the permitting for Stage II was scheduled to begin. This situation led to Edison proposing and the government agreeing to delay the permit applications until the latest practical time to see if circumstances were to turn more favorable for the continued operation of the Cool Water facility. Edison concluded by mid-1987 that Stage II would not be economical, and petitioned the government to release them from the obligation to continue the permitting effort. The government concluded, however, that energy prices would not have to increase dramatically for the project to be viable, and pressed Edison to leave the issue pending. Eventually, however, with continued low energy prices and a lack of support from the California Public Utility Commission, the Government agreed that Stage II would not be economic and released SCE from further obligation to operate the project. When the program ended, only $105 million of the obligated $120 million was used.

All in all, one could not have hoped for a more successful outcome of the Corporation's first selection of a project to assist. The price guarantee mechanism established by the Act functioned just as intended: the government assumed the financial risks that exceeded the means of the private sector, yet managed to leave all the key management decisions in the private sector so that they were made with maximum commercial and economic insight.

Dow Syngas

Initial construction of the Dow Syngas Project began in November 1984 and was completed within budget and on schedule in the spring of 1987. Production began in April 1987 and, according to the terms of the contract, the project was eligible for price guarantee payments the following month. After a learning period of shakedown, de-bottlenecking, and minor redesign, the project attained the contractually defined milestone of "Commencement of Commercial Production" on December 19, 1989.

The project encountered no significant operational difficulties during the ten-year period during which it was eligible for price guarantee payments. And, as is often the case for pioneer plants, years of operating experience resulted in improved on-line performance and production exceeding the original design parameters.

This led the project to request in its last year of operations, 1995, that the terms of the assistance agreement be modified. Because the contract limited the amount of price guarantee payments that the project could collect in any calendar quarter and because the plant was operating at higher levels for which Dow effectively could not collect guarantee differential payments, they proposed a change in the formula. In exchange for having the quarterly payment limits increased, they would agree to reducing the total amount of assistance made available in the contract: from $584,500,000 to $576,896,400. The government gained through the lower total outlays and Dow was able to save operating expenses as a result of ceasing production operations a year or so sooner than planned. In effect, the plant's flawless operation had already provided all the learning experience Dow had hoped to achieve.

As was the case for the other assisted projects, Dow had prepared an Environmental Monitoring Plan that was approved by a Monitoring Review Committee composed of members from the Departments of Treasury, Energy, and Defense, as well as the Environmental Protection Agency. The facility operated well within all permitted limits as was shown in quarterly and annual reports prepared by the project and reviewed by the committee.

As a result, the nation has another gasification technology proven at commercial scale upon which it can draw when energy market conditions make such action economical.

Great Plains

Chapter 11 described how the Corporation's proposed assistance agreement with Great Plains fell through because of opposition by Energy Secretary Herrington, at which time the sponsors promptly defaulted under the terms of the loan, as they had said they would do. As a consequence, DOE inherited the project because they were the guarantor of the loan. It was ironically appropriate for Secretary Herrington to have the responsibility for working out the project's further existence (or non-existence).

Under these circumstances, DOE hired an investment banker and sold the project to Great Basin in 1988 for a modest amount and the forgoing of the tax credits they would otherwise have gotten with the plant. This deal did not financially compare well with the last agreement that had been negotiated by the Corporation in which the sponsors would have put up an additional $190 million, forgone tax credits and would have repaid the $1.5 billion of loan for about $720 million in price guarantees to be spread out over a ten-year period. But Herrington's actions were obviously politically motivated to undermine the Corporation and had nothing to do with saving the government money, or trying to ensure long-term operation of the project.

Still, in the end, the project did continue to operate splendidly, and that was the important outcome. Indeed, over the next few years, the project turned into a modest moneymaker. The management sharply reduced staffing and operating costs and significantly improved the plant's capacity factor. So, with no debt burden and the continuation of favorable tariffs on the gas that had been granted by FERC, the project made roughly $100 million per year. (The pipeline companies challenged their continued obligation to pay the higher tariffs in court, but lost the case.)

Moreover, this technology can be considered commercially proven; indeed, at the time of this writing (2008), the plant continues to operate 24 years after completing construction, producing 54 billion standard cubic feet of natural gas annually.[176]

Parachute Creek

In late 1983, Unocal completed construction of the Parachute Creek project in 33 months (virtually on schedule). Getting the plant to operate was another matter. Indeed, Unocal's experience could have come directly from the generic conclusions of the Rand Study on pioneer plants discussed in Chapter 6, which had highlighted how scale-up of solids handling plants invariably encounter unexpected problems, whose resolution may take years. The reader may also recall that the SASOL facility in South Africa, which was using German technology developed in the Second World War, required about three years to produce near design levels.

Unocal immediately encountered such difficulties. Ironically, the problems did not center on the innovative rock pump, but on associated systems: the retort scraper, the shale shaft cooler, and the seal leg system. As part of these efforts, the plant underwent internal modifications starting in December 1984 continuing into 1985. But it was not until 1986 that the plant could be said to be operating and even then it was not for about a year and a half that it could be operated with predictable stability that would be required for ultimate economic operation. By the end of 1987, the project was generally on line for about 25 days a month. The next big push was to increase capacity. Design changes made in mid-1988 allowed the plant to operate at about two-thirds capacity, and slow improvements were made from there until mid-1991, when it achieved close to 75 percent capacity.

This level seemed to present an effective ceiling unless a further expensive design modification was made. While much of the Unocal experience is proprietary, they have disclosed the outline of the chief problems they confronted in getting the technology to work. As noted, the rock pump, which many observers found risky, operated without undue difficulty, while the "simple" spent shale cooling system presented the crucial difficulty.

Chapter 11 discussed how originally Unocal did not wish to install a fluidized-bed combustor as part of the "B" technology in order to limit the number of new systems being built for the first time—and thereby limit the risk. Accordingly, the spent shale was simply to be removed from the top of the retort and sprayed with water to cool it down before disposal. Alas, the non-sophisticated, *sure* approach contained the major problems. Initial analysis disclosed uneven distribution of cooling water that was easily corrected, but further attempts to operate were foiled by system instability at the bottom of the seal leg, where residual steam and entrapped gases are separated from the shale.[177] It required numerous modifications and many experimental modes of operation before the problem was resolved.

After seven years of effort that led Unocal to conclude they could bring the capacity up to design levels by making additional expensive modifications, the management did not feel that would be justified given low oil prices in the market and few prospects for a commercial synthetic fuels industry for a number of years. Moreover, even with the price guarantees the level of production capacity achieved resulted in negative cash flows, so that Unocal decided to cease operation of the facility altogether.

In additional to the technological learning experience, Unocal developed a wealth of data as part of the Environmental Monitoring Plan required under the contract. The project easily operated within permitted limits: air emissions were low, there was no discharge of water, and spent shale was re-vegetated so that there was no leaching of any contaminants.

That is how events unfolded for the Phase I Project. The reader may recall that the Corporation negotiated a contract amendment for an additional Phase II Project. The contract required Unocal to initially complete a detailed cost estimate for Phase II before proceeding further, and permitted Unocal to back out if the estimate proved too high. That is what occurred. When Unocal completed the cost estimate for the Augmentation Program in the spring of 1987, and found that the augmented facility would cost in excess of $350 million (vice the early estimate of $286 million), it opted not to proceed with the Augmentation Program. This action was contemplated by the Amendment and was carried out consistent with its terms. Consequently, the Treasury de-obligated the $500 million of authority that had been approved for the Augmentation Program. Thereupon, the project continued under the terms of the initial Phase I Agreement with its $400 million of price guarantee authority, modified only by some procedural provisions of the Amendment.

The wisdom of Unocal's dedication remains to be seen. But, if the nation needs large quantities of synthetic fuels from shale in the coming decades, Unocal will be the only entity having large-scale operating experience.

Forest Hill

Initial construction of the Forest Hill Project, involving buildings and roads, began at the site in November, 1984, and heavy construction was started in the last half of 1985. During 1986 all engineering and procurement efforts were completed. Consequently, the project was completed slightly ahead of schedule and under budget in the spring of 1987. Production began in April 1987 and the project was eligible for price guarantee payments the following month. The project went through a period of shakedown, de-bottlenecking, and minor redesign before attaining the milestone defined as "Commencement of Commercial Production" on December 19, 1988.

Although the project was constructed within budget and within the schedule initially set out, and while it demonstrated how to successfully deal with the major technological risks associated with the injection of high purity oxygen deep into the ground and how to control the resulting fire flood, it was not an economic success.

This was because of a number of operational problems that prevented the project from achieving production levels on the schedule that had been projected. One important problem was a delay in getting oxygen injected into the field because Union Carbide was late in completing the oxygen plant and then its operation was delayed until Carbide was convinced that its operation would be safe. (This was just after the Bhopal, India, tragedy that induced extra caution in chemical plants worldwide.) Delaying the oxygen and the associated combustion delayed, in turn, the build up of pressure and temperature in the reservoir and, thus, production of oil.

The project did well to reach the initial production milestone of 1,000 barrels per day for 45 days by April 1987, and rendering them eligible to receive price guarantee payments. And it pushed to increase production as fast as possible in order to accelerate cash flow. Unexpectedly, this approach presented unforeseen difficulties, not uncommon with new technologies. The problem was that production came disproportionately from a few wells. For example, at full field design level of 1,750 barrels per day, each of the 100 producing wells would have averaged about 17 barrels per day. During this period a few of the wells were "free flowing" (requiring little pumping) at a rate in excess of 100 barrels a day. These high rates did let the field reach an aggregate production in excess of 1,400 barrels per day, just short of achieving the "Commercial Production" milestone. But the ensuing wear and tear on the wells and the sanding that occurred raised maintenance costs to an unsupportable degree and often resulted in wells being placed out of service. It became obvious that the project had not achieved a viable mode of long-term operation.

Consequently, the project moved to a "balanced mode" of operation in which individual wells were not allowed to produce faster than consistent with reasonable costs. It was hoped that the lower producing wells would gradually increase their production as the outflow on the bigger producers was constrained. This phenomenon did begin to be observed. In the meanwhile, however, total field production temporarily fell back to 1,000 barrels a day for several months. Consequently, the project had cash flow reduced by several hundred thousand dollars per month, such that when the first repayment of principle on the guaranteed loan was due in December 1988, the project was short about $1.8 million to make the payment. The project's sponsors considered making it anyway, but they had already added several million dollars to the project over the prior two years to make up some of the earlier shortfalls. Given the low oil prices prevalent at the time, they appeared to believe that a further investment was not called for, and declined to make the contractually required payment. Accordingly, the government declared the project to be in default and repaid the bank the amount of the guaranteed loan. The Government also terminated the price guarantee payments and the Forest Hill Company filed for bankruptcy within two weeks of missing the loan payment.

This outcome would not have been unexpected to the framers of the Act, in that they expected the relatively risky synthetic fuels technologies to experience a number of failures. And much was learned. But, in the final analysis, the nation did not gain a proven technology for operating fields of this kind with oxygen fire-flooding technology. There were still too many uncertainties with regard to whether the field would have reached design levels of production and whether the "balanced mode" would have kept operating costs within acceptable bounds.

Nevertheless, environmental experience was gained in that the project has operated within the permitted conditions and had collected the monitoring data required by the EMP. As was true for other projects receiving assistance from the Corporation, a Monitoring Review Committee composed of the agencies mentioned elsewhere was established to review and assess results contained in quarterly and annual reports prepared by the project.

COST TO THE GOVERNMENT

The total costs of the synthetic fuels program associated with the Corporation[178] and follow-on administration by Treasury[179] were as shown in Table XIV-1.

This total mounts to not much more than one percent of the $88 billion authorized by Congress for the synthetic fuel effort. For completeness, however, to this figure should be added the $1.5 billion amount provided by the Department of Energy to the Great Plains Project. In any event the total is a small fraction of what Con-

SFC PROGRAM EXPENDITURES

PROJECT ASSISTANCE	AMOUNTS ($ millions)
Cool Water	105
Dow Syngas	576.9
Parachute Creek	134.2
Forest Hill	36.5
ADMINISTRATIVE EXPENSES	
SFC (1981-1986)	96.3
Treasury (1986-1996)	11.5
TOTAL	960.4

Table XIV-1

gress originally anticipated and much less than many contemporary commentators cite as the cost of the synthetic fuels program.

BENEFITS GAINED

For this expenditure, the country gained much in the way of technological development, environmental performance data, and understanding of the use of innovative forms of financial assistance. While these have been covered in the context of the project descriptions, a brief recapitulation follows.

Technology Development

In summary, the nation made considerable advances in its capability to draw on its vast energy resource base contained in oil shale and coal. Three different coal gasification technologies have been perfected from which a natural gas equivalent or gasoline (through indirect catalytic processes) can be produced. For oil shale, only one technology has been built and operated, and it still requires technological modification to attain design production levels. But Unocal's perseverance provided seven important years of experience in building a pioneer facility and working through most of its operating problems. The project also gained valuable experience by developing shale mining technology that was required for the excavation of caverns vastly larger than typical for coal or other conventional mining operations. Moreover, the project perfected upgrading technology for refining the crude shale oil that removed contaminants and produced high-grade sweet petroleum crude oil.

To be sure, it would have been desirable if the Basic Business Plan of the Corporation had been allowed to proceed to fund another couple of shale technologies to widen options available to industry. Nonetheless, the country is far better off, not having to start from scratch.

Environmental Performance

The Energy Security Act called for "the creation of commercial synthetic fuels production facilities …in an environmentally acceptable manner."[180] Further, Section 131(e) provided that:

> [a]ny contract for financial assistance shall require the development of a plan, acceptable to the Board of Directors, for the monitoring of environmental and health-related emissions from the construction and operation of the synthetic fuel project. Such plan shall be developed by the recipient of financial assistance after consultation with the Administrator of the Environmental Protection Agency, the Secretary of Energy, and appropriate State agencies.

Moreover, all the Corporation's solicitations required that each project identify all of the environmental permits and approvals required for project construction and operation along with a schedule for obtaining the same. The Corporation's environmental evaluation staff communicated with all relevant agencies to assure themselves that these requirements were being met. Each project prepared the required monitoring plan in consultation with the relevant agencies and prepared quarterly and annual reports that were reviewed by a Monitoring Review Committee, which had representatives from the federal agencies specified by the Act.

Years of monitoring demonstrated that the projects were environmentally benign in all respects. A report by the Electric Power Research Institute prepared at the conclusion of the Cool Water project's operation concluded.[181]

The plant demonstrated compliance with all permit-mandated emissions limitations, including the stringent requirements of the San Bernardino County Air Pollution Control District. When operating on either the low-sulfur western coal or the higher-sulfur eastern coals, 97% sulfur removal was achieved. The plant emitted not more than 15% of the Federal New Source Performance Standards (NSPS) level for NOx, 19% of the SO_2 NSPS level, and 33% of the particulate NSPS level for coal-fired plants. ... nitrous oxide (N_2O) emissions from the HRSG stack were also measured; levels were very low—0.5 ppmv.

As seen, the project easily met all emission limitations regarding air quality. Since the project was designed to have essentially no liquid discharge, it also met any such limitations. Wastewater was treated and reused, or it was impounded in lined evaporation ponds.

Elemental sulfur and slag were the two solid waste products produced by the plant. The elemental sulfur was sold under an annual contract, and the slag was temporarily stored in an impermeable lined pit. The California Department of Health Services ruled that the slag is non-hazardous. Such performance was characteristic of the other coal projects.

Shale oil production by Parachute Creek proved to be equally environmentally benign. Specifically, air emissions were well within permitting limits, there was no water discharge, and spent shale was disposed of in engineered, re-vegetated landfills that produced no leaching to the environment. Moreover, supplies of water and electricity presented no problem.

Copies of the Environmental Monitoring Plan Reports for all the projects are available through the National Technical Information Service (NTIS).

The professional environmental community had fiercely opposed the creation of a synthetic fuels industry by playing on speculative fears about the environmental impact of such facilities, fears
that could not at the time be convincingly refuted in the absence
of hard data. That data was collected and analyzed as a result of
the synthetic fuels program described in this book. The uncertainty
and unsubstantiated fears from the past can be put to rest.

It is of course unlikely that such fears will disappear altogether.
The opposition by the environmentalists never involved facts, but
was highly ideological. Undoubtedly, they will resort to attacks
based on global warming. But reality is such that the country will
require liquid mobility fuels for decades to come, and synthetic fuels are superior to other alternatives to petroleum such as biomass
in terms of economics, environmental impact, net energy consumption, or net CO_2 emissions.

Forms of Financial Assistance

As discussed in Chapter 4, the Energy Security Act provided for
a historically unique package of financial incentives to be used to
induce the private sector to begin to build a commercial synthetic
fuels industry. While the government could have built government
owned and managed facilities (as they did at the beginning of the
nuclear industry), a higher priority was placed on having the initiative in the private sector, given that it had the expertise and talents
necessary to insure that the plants were appropriately designed
and managed.

Congress was uncertain as to which financial instruments
would be most efficacious in meeting the goals of the Act and provided an array of such: price guarantees, loan guarantees, purchase
agreements, direct loans and joint venture assistance. The Corporation's efforts in negotiating the four assistance contracts as well as
nine letters of intent provided hard experience that can inform any
future endeavors in which the government needs to enlist the efforts of the private sector to build an industry. Indeed, a review of
the SFC's contract negotiations shows the power of price and loan
guarantees to motivate sponsors. These eliminate market risks for
the sponsor and helped them share completion risk. Moreover, the

innovative packaging of assistance crafted by the SFC financial staff resulted in minimizing the financial authority being obligated.

The actual experience of the four assisted projects dramatically underscores the utility of these forms of assistance from the perspective of the government. Most notably, while the Corporation obligated up to a total of $1.15 billion to the four projects, only $852.6 million was expended. The principal reason for the difference is that when Parachute Creek and Forest Hill ran into technical difficulty, they did not operate long enough to draw down the full value of the price guarantees awarded to them.

The specifics for Parachute Creek are particularly illuminating. When committing to proceed with the project on the basis of the $400 million price guarantee, Unocal estimated the cost of the facility to be about $450 million. Construction actually ran closer to $650 million. Then, if one were to factor in the costs of maintaining the facility through several years of startup attempts and continuing to make design modifications, the total tab was more on the order of $1.2 billion. The total tab to the government was only $134.2 million: quite a deal for the government.

IGNORANCE OF THIS HISTORY

Looking back over this saga at a time of renewed concern about sources of energy, is it not strange how little of it is generally known? As this is written in 2008, the ignorance is palpable. Wildly incorrect references have begun to appear as more and more pieces are published. For example, one article in *The National Review* asserted, without qualification that the Corporation had funded only two projects, both of which were total failures. Senator Ken Salazar (D-Colorado) stated in an op-ed piece regarding shale oil technology: "past efforts have failed miserably," that there was a "failure to plan for environmental and social impacts," and that we do not know how much water is needed, how much carbon would be emitted or what the effects would be on Western landscapes. (*Washington Post*, July 15, 2008) Clearly, he was clueless about the experience of the Parachute Creek Shale Oil Shale Project.

Another reference in a *Wall Street Journal* editorial was more egregious: "The Carter-era Synthetic Fuels Corporation—one of the

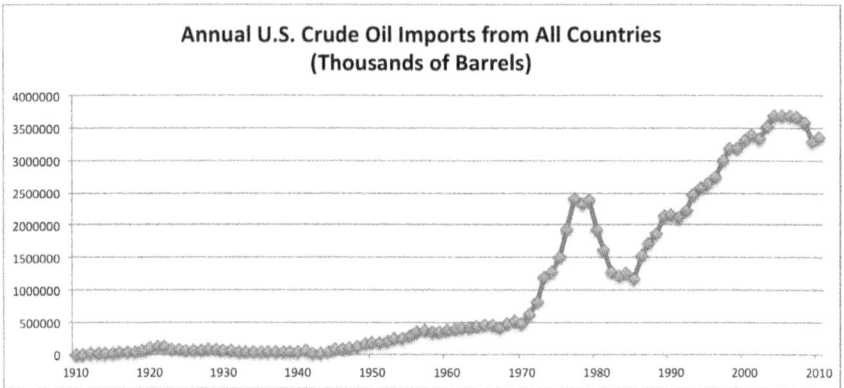

Figure XIV-1

more notorious Washington boondoggles of all time, ... spent $2.1 billion of tax dollars on alternative fuels before declaring bankruptcy." (*Wall Street Journal*, January 16, 2007) As the reader by now can easily ascertain, the SFC did not declare bankruptcy, did not spend $2.1 billion, and provided substantial technical learning.

More importantly, as a result the country is better positioned to make economic use of its coal and oil shale resources as its imports of natural petroleum continues to increase (see Figure XIV-1) and the real price of crude continues to climb in tandem with global economic growth—as projected by the SFC's Comprehensive Strategy Report twenty years earlier.

LOOKING AHEAD

Beyond the technology development and environmental learning that resulted from the Energy Security Act and the experience of the SFC, what other lessons might we take away as inevitable future energy challenges confront the United States?

Foremost, keep economics in the forefront of new policy initiatives. Energy resources are exhaustible commodities subject to inexorable laws of supply and demand. Because modern economies and growing standards of living globally will require increasing amounts of energy—the lifeblood of rising productivity—energy prices will rise in tandem bringing new, currently too expensive resources on line. We have already witnessed the emergence of a new tar sands industry in Canada and are now seeing an unexpected source of gas from shale in the United States. Synthetic fuels may or may not come into their own. But Biofuels and green technologies are not likely to be cost competitive, even if sentimental favorites. Let the market decide.

This saga can be viewed as a caution for Congress to show more humility before attempting to reshape a manifold and highly complex part of the economy, and for proponents of such measures to reflect that what Congress authorizes in new endeavors, it may renege on when public enthusiasm wanes and costs become evident.

Appendix A

Primary and Secondary Coal Conversion Technologies

PRIMARY AND SECONDARY COAL CONVERSION TECHNOLOGIES

Primary Conversion Technology	Secondary Conversion Technology	End Products
Gasification	None	Industrial heating or power generation fuel
Indirect Conversion	Methanation	Methane (pipeline or high-Btu gas)
	Methanol synthesis	Methanol (chemical grade, methyl fuel)
	Methanol synthesis plus Mobil-M catalytic synthesis	Gasoline
	Fischer-Tropsch catalytic reaction	Range of products including: * Gasoline components * Synthetic Crude * Chemicals * Methane gas
	Shift	Hydrogen
	Ammonia synthesis	Ammonia
Direct liquefaction (possibly preceded by low-temperature carbonization)	Petroleum refinery type processes	* Synthetic crude * Gasoline * Fuel oil * Jet fuel * Kerosene * LPG

Appendix B

Solicitation Summary

Solicitation	Date Issued	Proposal Submission Deadline	Proposals			
			Submitted	Rolled Over	New	Resubmittals
Initial General	11/21/80	3/31/81	68	n/a	68	n/a
Second General	12/11/81	6/1/82	35	3	14	24
Third General	8/19/82	1/10/83	46	3	17	32
Western Oil Shale	1/20/83	3/15/83	6	n/a	1	5
Gulf Lignite	4/25/83	7/25/83	1	n/a	1	0
Eastern Coal	6/30/83	12/1/83	9	n/a	7	2
Coal/Lignite	1/5/84	2/2/84	1	n/a	0	1
Coal - Water	1/5/84	6/15/84	11	n/a	11	n/a
Retrofit	2/16/84	6/21/84	2	n/a	1	1
Fourth General	2/16/84	6/29/84	18	n/a	5	13
Eastern/Int. Bituminous	5/28/85	7/8/85	10	n/a	8	2
Tar Sands	6/24/85	8/30/85	4	n/a	2	2

Appendix C

EXHIBIT C-1

TIMELINE OF KEY SFC EVENTS

	1980	1981	1982	1983
LEGISLATIVE/ GOVERNANCE	Energy First Board Security Board Resigns Act Apptd	New Board Confirmed	SFC Operational	
SOLICITATION ACTIONS	Initial	2nd / Mi Si	3rd / M2 S2	Tos Tgl Teic / M3 SJ Mos Mgl Meib
LETTERS OF INTENT			Santa Rosa | Cool Water | Seep Ridge / First Colony | Cathedral Bluffs / Calsyn | Parachute Creek II	
ASSISTANCE CONTRACTS	Parachute Creek (DOE) Colony (DOE) Great Plains (DOE)			First Cool Colony Water (Design) PGC

Legend: Txx (Issuance of Targeted Solicitation): Oil Shale (os), Gulf Lignite (gl), Eastern/ Interior Coal (eic), Coal/ Lignite Gasification (clg), Coal-water Fuels (cwf), Coal lignite Gasification/ Retrofit (clgr), Eastern/ Interior Bituminous Coal (eib), and Tar Sand (ts).
Mi (Maturity/ Qualification Determinations: Initial, 2nd, 3rd, 4th, and targeted solicitations.)
Si (Project Strength Determinations: Initial, 2nd, 3rd, 4th, and targeted solicitations.)

EXHIBIT C-2

TIMELINE OF KEY SFC EVENTS

	1984	1985	1986
LEGISLATIVE/ GOVERNANCE	Board $2b $5.8b New Loses Resc Resc Board Quorum	Comprehensive Strategy Report To Congress	Congress SFC Acts to Terminated Terminate SFC
SOLICITATION ACTIONS	Tclg Tclgr Teibc Tts Tcwf 4th Gen. Seib Melg	Mcwf Mcgr	
LETTERS OF INTENT	Dow Syngas Forest Hill Northern Peat Energy Kentucky Tar Sand HOP Kern River Great Plains		
ASSISTANCE CONTRACTS	Dow Syngas PGC	Forest Hill Agreement Parachute Creek II PGC	

LEGEND: Txx (Issuance of Targeted Solicitation): Oil Shale (os), Gulf Lignite (gl), Eastern/ Interior Coal (eic), Coal/ Lignite Gasification (clg), Coal-water Fuels (cwf), Coal lignite Gasification/ Retrofit (clgr), Eastern/ Interior Bituminous Coal (eib), and Tar Sand (ts).
Mi (Project Maturity/ Qualification Determinations: Initial, 2nd, 3rd, 4th, and targeted solicitations)
Si (Project Strength Determinations: Initial, 2nd, 3rd, 4th, and targeted solicitations)

Appendix D

Findings of the Hagler, Bailly Report[183]

In developing its findings concerning learning benefits from the operation of new processing technologies, the Hagler, Bailly Report examined the experience of three industries having significant relevance for synthetic fuels, i.e., low-density polyethylene (LDPE), ammonia, and alumina refining industries, which had special relevance for synthetic fuels production facilities because of their large solids-handling component, chemical synthesis, and high capital intensity. Some of the findings of the study included:

In LDPE manufacturing the principal feedstock is ethylene. Increased experience with the manufacturing process over nearly forty years (from 1943 to 1981) resulted in ethylene consumption falling from 1.33 pounds to 1.05 pounds per pound of LDPE. Over the same period, plant size grew from roughly 1 million pounds of LDPE per year to 550 million pounds of LDPE per year. As a result, investment requirements per pound of annual capacity decreased from $10.86 in 1943 to $0.13 in 1981 (in constant 1968 dollars).

The ammonia industry experienced a similar reduction in costs, although more of the reduction was relatively due to technological change (though the scale increased as well). Many of the technological improvements dealt with providing the hydrogen feedstock and coping with the high pressures of ammonia synthesis, which were reduced over time by nearly a factor of ten. The overall progress resulted from an array of incremental improvements in process technology, in the mechanical engineering of process components, in the integration of the process units, in the plant energy balance, and in the compressor drives. The net result was that by 1970 investment cost per ton of ammonia production capacity fell to 5 percent of the level required sixty years earlier (in constant dollars).

In the alumina industry a single technology, the Bayer process, has dominated. While the significant cost reductions that occurred can largely be attributed to increased scale, technology progress contributed as well. Plant size increased from 75 tpd to 1,000 tpd by 1950. The progress in technology involved a movement from rotary kilns to fluid-bed calciners that use one-third less energy, are less expensive to install, and require less maintenance.

Appendix E

The Santa Rosa Project Hearing

This Appendix contains more detail to augment that contained in Chapter 12 on the testimony covered in the hearing held by The House of Representatives Environment, Energy, and Natural Resources Subcommittee of the Committee on Government Operations on December 8, 1983.

There was seemingly little basis to hold a hearing on the subject of the Santa Rosa Project since no federal funds were expended. Consequently, the Committee's staff had to identify other plausible public interests to justify holding a hearing as they did regarding four hypothetical considerations. First, if an assumption underlying a decision was eventually proven wrong, someone's judgment, or at least the process, must have been flawed. Second, if the Corporation was devoting its effort to a project that ultimately was rejected, there must have been opportunity costs associated with better projects that lacked Corporation attention. Third, since the letter of intent was designed to help the sponsors increase their equity, the Corporation could be accused of helping raise money for a project with flaws. Fourth, might there have been shortfalls in the Corporation's review process regarding environmental matters. This played out as follows:

1. After issuing the letter of intent, it became clear that the specific resource base for the project was deficient in that there were many geologic discontinuities in the sands, such that the chemical assay method that had been employed gave misleading results, even though it was applicable to richer tar sands. The Committee called on testimony from Tosco Corporation (which ultimately employed the more appropriate assay techniques for Foster-Wheeler)

and from Len Axelrod, the Corporation's Vice President for Technology and Engineering. The Committee hoped to show that the SFC's learning of the problems with the resource after the letter of intent was issued was a failure of the Corporation's evaluation process or a sign of the project's venality. As made clear in Chairman Noble's testimony and in Len Axelrod's responses, the simple answer is that all complex projects must address large uncertainties; getting information to reduce uncertainty takes time and money; that process must proceed in a deliberate and sequential manner; the cost, if possible, should be laid on the private sponsor; and all this was done and worked as intended.

The following summarizes the course of events:

The SFC staff recommended to the Board that the Santa Rosa Project be advanced to the Phase II evaluation status provided that the project, among other things, operate the integrated pilot plant for a satisfactory period (90 days), and it provide confirmation of the resource appraisal.

The SFC engaged the Laramie Energy Technology Center to provide it with an independent appraisal of the project's intended resource. The Center's Report stated that "We have estimated the lease contents at 29 million barrels of tar in the ore that contains 4% or more tar by weight. That estimate is 15% lower than the estimate made by the New Mexico Institute of Mining and Technology [whose study had been relied upon by the project]. We are not highly confident of this estimate because too few samples were analyzed for tar content. A sufficient number of wells were cored to provide a highly reliable estimate, but too few core samples were analyzed..." This amount of tar was more than enough for the intended project and the project had additional lands under lease as well.

The letter of intent signed by the Corporation had as two of its conditions: the characteristics of the bitumen to be produced by the project shall have been verified to the satisfaction of the SFC, and the partners or affiliates shall have operated an integrated pilot plant for at least 90 days (including a run of at least 21 consecutive days) and confirm the technical and financial feasibility of the project. The latter condition would entail the beginning of mining at the site.

Foster-Wheeler, who would be providing most of the equity for the project, quickly undertook two key efforts. First, they had an additional 19 core holes drilled. Then, they engaged Tosco Corporation to conduct a core drilling program (of an additional five holes), as well as laboratory analyses for the Santa Rosa Project site.

These efforts disclosed hitherto unsuspected and eventually insurmountable problems with the resources at the proposed site. Tosco employed two chemical assay methods to determine the richness of the resource – i.e., (1) ashing to determine the bitumen content by weight loss, and (2) organic solvent extraction techniques to determine the bitumen content. The former had been the basis for prior estimates at the site. It was relatively easier and was applicable to richer resources, but was misleading on leaner resources such as at Santa Rosa, particularly where the resource contains carbonates which decompose during the ignition process to release carbon dioxide. Thus, the resource turned out to be leaner than anticipated by a crucial two or so percentage points. Moreover, the resource was shown to be badly faulted, namely one could not assume continuity of resource between the core holes. Quite lean sands were found between two earlier core holes showing richer deposits.

The project's lease included substantial lands extending beyond the area originally intended for the mine. Consequently, the project investigated the resource in other parts of the lease, but determined that they were no more promising.

Foster-Wheeler withdrew from the project. Thereafter, the SFC withdrew the letter of intent at no cost to the government.

2. In an attempt to find an "environmental" issue, the Committee called as a witness Mr. Robert M. Findling, Director of Planning and Development, the New Mexico Natural Resources Department, State Park and Recreation Division (hardly a high-level spokesman for the State). The essence of his testimony was that his division had not been contacted by either the project or the SFC, that no Environmental Impact Statement was required of the project, and that he felt that there were unanswered questions – about which he proceeded to speculate. He noted that the project would be adjacent to a State Park serving recreational needs, and that the aesthetics of the entrance to the park might be affected by the pres-

ence of the mine. Otherwise, he questioned whether air quality would be impacted by wind erosion, whether inundation of mine pits would cause leaching of chemicals into reservoirs, and he noted that the discharge permit was incomplete. Synar did his best to play up the uncertainty of knowledge and the lack of consultation with the State. Unfortunately, his staff had provided him with a weak reed for this purpose.

First, Mr. Findling was grilled by Congressman Williams, who was friendly to the Corporation. Then, Steven Gottlieb, Director of Environment for the SFC made a number of telling points: the State Park had been in existence only two years following a court decision making it clear that the mineral rights were open for lease and Mr. Findling conceded that he had no specific knowledge of environmental problems; his points were merely hypothetical.

Further, the SFC had met with state officials of permitting agencies at length (the State Park and Recreation Division had no permitting authority over the proposed project). Moreover, the SFC had held public input meetings that were advertised in the locality to which no member of the State Department of Natural Resources attended. Mr. Findling had to acknowledge that the project was strongly supported by the Governor. At this juncture, Chairman Synar had to interject, "Mr. Findling doesn't sit here today to represent the Governor or anyone else."

Finally, Mr. Gottlieb noted that (1) no project receiving assistance from the Corporation was required to do an EIS simply because of receiving the assistance and, further, this project did not have to do an EIS under the Corps of Engineers regulations because the project would have been above the mean high water mark (i.e. any implication that the project or the Corporation was trying to evade an EIS or permit requirement was false); (2) the project was subject to air, water, and reclamation permits, which any project would be subject to; and (3) an Environmental Monitoring Plan had been drafted and was undergoing revision subject to comments of the participating agencies.

3.   During the period that the project was undertaking to meet the conditions of the letter of intent and some of the questions with regard to the resource base were arising, Solv-Ex was raising addi-

tional capital with the help of Morgan Stanley. Synar did his best to insinuate that Solv-Ex knew about the inadequacy of the resources while it was raising equity and somehow had misled the investors. After grilling John Rendall (project sponsor) and obtaining a letter from Morgan Stanley, it was clear that there was no case. The equity solicitation was conducted in June of 1983, and completed in July, but the knowledge of the resource was nailed down through the period of June through August.

Appendix F

A Representative Letter of Inquiry from Congressman Dingell

The following questions were taken from letter dated July 18, 1983 from Congressman Dingell to the SFC that was representative of many such. The author has included a response in brackets in some instances.

1. "What percent of equity is being put up by the First Colony Project Sponsors? Please list the percentage of each. What rate of return on equity will each sponsor obtain?" [*The details of the sponsors' participation were included in the term sheet summaries, which had already been issued to the public and the Committee. The eventual rate of return was problematical, depending largely on the sponsors' success in building and operating the facility. If all went according to plan, the SFC staff estimated a 16.5 percent rate of return.*]

2. "Is the $4.65 million committed by the SFC for 49% of the design to be repaid by the sponsors or is this a non-recoverable federal contribution?"

3. "Explain the requirement for assigning $820,750 to refine cost estimates. Is this a normal practice in the awarding of contracts? How much money is First Colony contributing to the study?" [*Section 131(u) of the Energy Security Act provided for and defined the requirements for the use of design cost sharing funds. The project was putting up 51% of the total.*]

4. "Is the First Colony Project fully committed to production at design production rates? If not, please explain the commitment." [*What else? If the project didn't perform at these rates, commensurately less price guarantee payments would be paid, and the sponsors would not receive the intended rate of return. But, if the pioneer technology didn't function at design rates, the sponsors could not be forced to achieve them.*]

5. "Is the current market for the use of methanol as a blending agent or fuel extender greater than available production? What studies do you have to support your conclusion?" [*Of course not—the U.S. hadn't had rationing since the Second World War. The point of the Act was to be displacing imported oil, which would be the indirect effect.*]

6. "The First Colony Project would appear to be in competition with methanol produced by the Getty refinery in Delaware. How do you justify government subsidies to a project that will compete with a private-sector financed development?" [*Every project funded by the Corporation would be in competition with a private-sector financed development somewhere or other.*]

7. "Does the SFC have a strategy developed for funding alternative resources such that those with the greatest potential are funded first with lesser potential following in rank order? If so, please list the technologies in rank order." [*The Corporation had completed in August, 1982, a reasoned and detailed statement of its priorities in funding different technologies to achieve the diversity objective of the Act. The Committee, of course, had a copy of it. The purpose of the question was to attack peat, which was not at the top of the list. Nonetheless, peat was explicitly listed in the Act as eligible, and the Corporation was not going to put off dealing with sponsors willing to commit substantial equity until all the top priority funding was accomplished.*]

8. "Does the SFC plan to recommend changes to the ESA mandated requirement for 500,000 BPD by 1987? If so, describe the changes." [*The Corporation was not about to answer this question. The Comprehensive Strategy required by the Act was not due until a year later.*]

9. "What is the projected price per barrel of natural gas and oil over the 1987-91 time frame?"

10. "Assuming that the SFC is following a policy of loan guarantees and price subsidies, what is the economic trade off between this policy and the alternative of GOCO development?" [*The Act required the SFC to place highest priority on the use of loan and price guarantees, and to fund projects through the Government-owned, Contractor-operated mechanism as a last resort to gain a desirable technology. The SFC Board had no intention of funding GOCOs.*]

11. "What is the value of the tax benefits per barrel as addressed in the Minutes of the SFC Board Meeting of December 1, 1982?"

12. "What SFC solicitation procedure allowed for the insertion of the First Colony project in the Second Solicitation after it had closed? Please provide a detailed explanation as to why this action was proper. What type of precedent does this set for other applications?" [*The solicitation explicitly provided for the eventuality that a project being considered under the first solicitation might be dropped after the closing date for the second solicitation. It required that the sponsors simply write the Board with a request that in that eventuality the project could be considered competitively in the second solicitation without incurring the expense of submitting another detailed proposal. The First Colony Project had done so.*]

13. "What material was presented at the December 2, 1982 Board Meeting that gave the Board significant confidence in the project such that SFC assistance could be authorized?" [*The Board had, of course, all along received detailed reports evaluating the project in the Maturity Reviews, the Strength Evaluations, and the results of the negotiating process as the Term Sheet was being developed. More specifically, the SFC had provided Dingell all the materials presented to the Board at that December meeting as Appendix 20 to the February 9, 1983, letter in which Noble had responded to earlier questions. As noted earlier, there was little indication that the SFC responses - laboriously prepared - were ever read in the same spirit.*]

14. "When the SFC Board voted to not give further consideration to the First Colony Project, the application was for a price guarantee. When approved by the Board, the application included a loan guarantee. What is the basis for a higher degree of participation by the SFC in this project? Would the sponsors continue to advance this project without the loan guarantee?" [*There was nothing binding on the projects in terms of the initial proposal outlines of financial assistance. This was to be determined through the competitive negotiating process. In this instance the sponsors had an early indication from a bank indicating that non-guaranteed financing could be arranged. When it came time to firm a commitment, however, the bank was unwilling to follow through. This likely eventuality was one reason that the Congress included the loan guarantee mechanism in the Act.*]

15. "What "appropriate incentives", as referred to in the "First Colony Project Phase IIA Findings", were added to improve the project's "marginally viable" economic condition?"[*See the prior question.*]

16(a). "The technical evaluation of the KBW gasifiers indicates that if the equipment will work on peat it will work on coal. Considering the limited potential of peat resources in the U.S., why wouldn't it be more economically attractive to demonstrate the process on coal, whose resource base is enormous?" [*The other three projects that had applied to the SFC for funding to build these gasifiers using coal as a feedstock were unable to attract sufficient equity to proceed. First Colony was successful - i.e. in effect, the market determined that it made most economic sense to build the first of these gasifiers using peat rather than coal as a feedstock. Moreover, First Colony had agreed to test coal in its gasifier.*]

16(b). "What are the potential peat resources in the U.S. for the conversion to methanol? In your reply, please consider the several peat resource contracts entered into by the Energy Department."

17. "Was there any change in the sponsorship of the project between the time the project was dropped in June and the time it was approved in December? If so, who were the sponsors that were added or deleted and why?" [*None were dropped. Transco was added.*]

18. "Did any of the sponsors of the First Colony project have direct contact with any members of the SFC between June and December 1982? If so, list the names, dates and substance of the contacts." [*This could not be a serious question. The evaluation and negotiation processes required extensive day-long interactions. Appendix G to the SFC response to these questions provided the desired information.*]

19. "What impact will run off from the First Colony peat fields have on the commercial fishing industry in North Carolina?" [*None. The project was designed to have zero incremental runoff.*]

20. "What qualifying conditions remain to be met before the project is fully committed to design production capacity?" [*See question 4.*]

21. "What was the result of the gT/HR test unit?" [*A pilot plant test of the North Carolina peat.*]

22. "Please explain the data in the unedited version of "Summary of Project Evaluation Results" for the First Colony Farms in

the table identified as "Ideal and Specific Sponsor - Market Value of Assistance" and indicate why it should be confidential." [*The SFC had indicated to the Committee staff what portion of materials previously supplied were sensitive and should not be freely released. In this instance, the sponsors had provided capital and production cost data that was proprietary.*]

23. "Does the technical data submitted for the First Colony Project indicate that environmental considerations have no restraining potential on plant production?" [*Yes.*]

24. "The ESA intended that financial assistance be extended to technologies which offer significant potential for use as a synthetic fuel feedstock as well as offer the potential for achieving the national synthetic fuel production goal. Provide the statistical data which indicates that the First Colony Project meets both of these criteria." [*Provided in Appendix F of the SFC response.*]

25. "In estimating the total national potential of peat resources, what percent has been discounted as reasonably available as a result of environmental limitations in Alaska and the Northern Tier?"

Works Cited

Note: Unpublished documents of the U.S. Synthetic Fuels Corporation were archived at the U.S. General Records Center.

Anderson, Jack. "Synfuel Hunters Explore Saunas and Nightclubs", *The Washington Post*, August 23, 1983.

Churchill, Winston. *The Gathering Storm*. New York: Houghton Mifflin Harcourt, 1986.

D.C. Duncan and V. E. Swanson, "Organic-Rich Shales of the United States and World Land Areas." *USGS Circular 523*, 1965.

Dakota Gasification Company Internet Site, September 24, 2008.

Daily News, "Monks, Cohen Allied in Synfuels Power Struggle" (Bangor Maine).

Electric Power Research Institute, *Cool Water Coal Gasification Program: Final Report* (EPRI GS-6806, Project 1459), December 1990.

Executive Office of the President, Office of Management and Budget, Letter from James C. Miller III, Director, to the Honorable Mark O. Hatfield, United States Senate, December 12, 1985.

Environmental Policy Institute, *Dreams into Dollars: A Review of Five Candidates for Synthetic Fuels Corporation Assistance*, 1982.

Environmental Policy Institute, *Distant Dreams: The Synthetic Fuels Corporation's Search for a Commercial Industry*, 1982.

Hagler, Bailly and Co. *Learning Curves: Empirical Evidence of Production Cost Improvement*, HBC Ref. No. RA85-420-2; Washington, D.C., June 1985.

Hawken, Paul, Ogilvey, James, and Schwartz, Peter. *Seven Tomorrows*. 1982.

Herrington, John. Op-ed article, "The Synfuels Energy Dinosaur". *The Wall Street Journal*, October 9, 1985, New York Edition.

_____. Letter to Edward E. Nobel, May 21, 1985.

Hess, R.W. *Potential Production Cost Benefit of Constructing and Operating First-of-a-Kind Synthetic Fuel Plants*, WD-2577-SFC. The Rand Corporation, March, 1985.

Knowles, Ruth Sheldon. *America's Energy Famine: Its Cause and Cure.* Norman: University of Oklahoma Press, 1980.

Lipkin, Jeffrey. E-mail to Hervey Priddy, February 24, 2011.

Manchester, William. *The Last Lion: Winston Spencer Churchill ALONE 1932-1940.* Boston: Little, Brown, and Company, 1988.

McDermott, Geoffrey A. "The Energy Security Act of 1980", July 6, 1987.

Merrow, Edward W., Phillips, Kenneth E. and Myers, Christopher W. *Understanding Cost Growth and Performance Shortfalls in Pioneer Process Plants.* The Rand Corporation, September 1981.

Miller, James C. III, Letter to Honorable Mark Hatfield, U.S. Senator, December 12, 1985.

William F. Rhatican, Letter to Jack Anderson, August 25, 1983.

Schumacher, E.F. *Small is Beautiful: Economics as if People Mattered.* Canada: Harper Collins Publishers, 1974.

Schurr, Sam H. and Netschert, Bruce C. et.al. *Energy in the American Economy, 1850-1975, Its History and Prospect.* Baltimore: The Johns Hopkins Press, 1960.

Sobel, Lester A. (ed.). *Energy Crisis: Volume 1, 1969 – 73.* New York: Facts on File, Inc., 1974.

_____. *Volume 2 (1974-1975),* New York: Facts on File, 1975.

_____. *Volume 3 (1975-1977),* New York: Facts on File. 1978.

_____. *Volume 4(1977-1979),* New York: Facts on File, 1980.

Stobaugh, Robert. *Energy Future.* New York: Ballantine Books, 1980.

Thompson, V. M., Jr.. Letter to the Board of Directors of the U.S. Synthetic Fuels Corporation, April 26, 1984.

U.S. Congressional Record - House, July 25, 1984. Remarks by Congressmen Conte, Dingell, Synar, Tauke, Wolpe, and Wright.

U.S. Congressional Record – Senate, September 26, 1984, and October 3, 1984. Remarks by Senator McClure.

U.S. Congressional Record - Senate, October 31, 1985. Remarks by Senators McClure, Johnston, and Metzenbaum.

U.S. Public Law 96-294, The "Energy Security Act", 94 Stat. 611, including the 1980 Amendments to the Defense Production Act of 1950"

U.S. Energy Administration, "Annual U.S. Crude Oil Imports from All Countries," http://tonto.eia.gov., October 15, 2008.

U.S. Public Law 98-473, "Making Continuing Appropriations for the Fiscal Year 1985, and for Other Purposes."

U.S. Public Law 99-190, Further Continuing Appropriations for Fiscal Year 1986.

U.S. Congress Hearings:

U.S. House of Representatives, Committee on Government Operations/ Subcommittee on Environment, Energy and Natural Resources. Hearing on the Synthetic Fuels Corporation. 97th Cong., 1st Sess., February 14, 1981.

U.S. House of Representatives, Committee on Energy Research and Development/ Senate Energy and Natural Resources Committee. Hearings on the Synthetic Fuels Corporation. 97th Cong., 1st Sess., April 3, 1981.

U.S. House of Representatives, Committee on Energy and Commerce/ Subcommittee on Fossil and Synthetic Fuels. Hearing on Synthetic Fuels Policy. 97th Cong., 1st Sess., July 9, 1981.

U.S. House of Representatives, Committee on Science and Technology/ Subcommittee on Energy Development and Applications. Hearing on "Synthetic Fuels Development". 97th Cong., 1st Sess., July 27, 1981.

U.S. House of Representatives, Committee on Government Operations/ Subcommittee on Environment, Energy and Natural Resources. Hearing on "Oversight: Goals of the Reagan Board of Directors". 97th Cong., 1st Sess., September 17th, 1981.

U.S. House of Representatives, Committee on Science and Technology/ Subcommittee on Energy Development and Applications. Hearing "Synthetic Fuels Environmental R&D". 97th Cong., 1st Sess., October 1, 1981.

U.S. House of Representatives, Committee on Energy and Commerce/ Subcommittee on Fossil and Synthetic Fuels. Hearing on the "SFC". 97th Cong., 2nd Sess., April 2, 1982.

U.S. House of Representatives, Committee on Science and Technology/ Subcommittee on Energy Development and Applications. Hearing on "Socioeconomic Impacts of Synthetic Fuels". 97th Cong., 2nd Sess., April 2, 1982.

U.S. House of Representatives, Committee on Government Operations/ Subcommittee on Environment, Energy and Natural Resources. Hearing on "Synthetic Fuels Industry in Today's Economic Climate". 97th Cong., 2nd Sess., June 9, 1982.

U.S. House of Representatives, Committee on Banking, Finance and Urban Affairs/ Subcommittee on Economic Stabilization. Hearing on SFC Progress. 98th Congress, 1st Sess., May 12, 1983.

U.S. Senate, Committee on Government Affairs/ Subcommittee on Oversight of Government Management. Hearing on SFC Management. 98th Cong., 1st Sess., July 27, 1983, and July 29, 1983.

U.S. House of Representatives, Committee on Energy and Commerce/ Subcommittee on Fossil and Synthetic Fuels. Hearings on Synthetic Fuels Policy. 98th Cong., 1st Sess., October 4, 1983, October 5, 1983, January 25, 1984, and June 18, 1984.

U.S. House of Representatives, Committee on Government Operations/ Subcommittee on Environment, Energy and Natural Resources. Hearing on the Great Plains Coal Gasification Project's application for assistance. 98th Cong., 1st Sess., October 18, 1983.

U.S. House of Representative, Committee on Government Operations/ Subcommittee on Environment, Energy and Natural Resources. Hearing regarding an "Examination of Procedures by which the SFC Selects Projects for Federal Financial Assistance". 98th Cong., 1st Sess., December 8, 1983.

U.S. House of Representatives, Committee on Energy and Commerce/ Subcommittee on Oversight and Investigations. Hearings regarding SFC President Victor Thompson. 98th Cong., 2nd Sess., April 3, 1984, and June 27, 1984.

U.S. House of Representatives, Committee on Government Operations/ Subcommittee on Environment, Energy and Natural Resources. Hearing regarding "SFC Oversight". 98th Cong., 2nd Sess., May 16, 1984.

U.S. House of Representatives, Committee on Science and Technology/ Subcommittee on Energy Development and Applications. Hearings on "The Status of Synthetic Fuels and Cost-Shared Energy R&D Facilities". 98th Cong., 2nd Sess., June 6, 1984, June 7, 1984, June 13, 1984.

U.S. House of Representatives, Committee on Government Operations/ Subcommittee on Environment, Energy and Natural Resources. Hearing on the proposed award of assistance to the Great Plains Project. 99th Cong., 1st Sess., May 22, 1985.

U.S. House of Representatives, Subcommittee on Oversight and Investigations of the Committee on Energy and Commerce, Letter from John D. Dingell and James T. Broyhill to Edward E. Noble, February 16, 1984.

U.S. Senate, Committee on Energy and Natural Resources, Letter from the Chairman James A. McClure to Colleagues, October 4, 1985.

Congress of the United States, House of Representatives, Letter from Member of Congress Michael L. Strang to the Editor of the *Wall Street Journal*, October 18, 1985. (Author's Files.)

U.S. General Accounting Office:
Special Care Needed in Selecting Projects for the Alternative Fuels Program, December 8, 1980.
Synthetic Fuels Corporation's Management of Demonstration Projects Would be Limited, July 10, 1981.
The United States Synthetic Fuels Corporation's Project Selection Guidelines Need Clarification, August 5, 1981.
Evaluation of Administrative Procedures at he Synthetic Fuels Corporation, October 18, 1982.

Environmental and Socioeconomic Status of the Hampshire Energy Project, October 22, 1982.

Synthetic Fuels Corporation's Use of an Expert Panel and a Staff Assistance Agreement, February 2, 1983.

Review of the United States Synthetic Fuels Corporation's Financial Statements for the Year Ended September 30, 1982, April 12, 1983.

Circumstances Surrounding the First Colony Peat-to-Methanol Project, November 20, 1983.

Letter from the Comptroller General of the United States to the Honorable Tom Corcoran re the legality of the United States Synthetic Fuels Corporation providing price guarantees to Great Plains Gasification Associates, January 19, 1984.

The U.S. Synthetic Fuels Corporation's Contracting with Individual Consultants, February 7, 1984.

Letter from the Comptroller General of the United States to the Honorable John D. Dingell re whether the United States Synthetic Fuels Corporation is adhering to the restrictions on financial assistance placed on it by the Energy Security Act, February 15, 1984.

Federal Efforts to Control the Environmental and Health Effects of Synthetic Fuels Development, March 9, 1984.

The Synthetic Fuels Corporation's Progress in Aiding Synthetic Fuels Development, July 11, 1984.

Plans to Award Additional Financial Assistance to the Union Oil Company Oil Shale Program, July 23, 1984.

Procedures Need Strengthening in the U.S. Synthetic Fuels Corporation's Conflict of Interest Program, September 20, 1984.

Financial Status of the Great Plains Coal Gasification Project, February 21, 1985.

Synthetic Fuels Corporation's Profit-Sharing Provisions with Six Proposed Projects, July 10, 1985.

U.S. Synthetic Fuels Corporation's Contracting Policies and Practices for Consulting Services, August 15, 1985.

U.S. Synthetic Fuels Corporation, Annual Reports, 1981, 1982, 1983, 1984, and 1985.

U.S. Synthetic Fuels Corporation, Analysis of Administrative Issues, April 24, 1985. (Author's Files.)

U.S. Synthetic Fuels Corporation Legal Services Group, Memorandum for the Board of Directors, "Victor   M. Thompson, Jr./ Utica Bankshares Corporation," April 23, 1984.

U.S. Synthetic Fuels Corporation, Letter from Edward E. Noble to the Editor of the *Wall Street Journal*, October 9, 1985. (Author's Files.)

U.S. Synthetic Fuels Corporation, *Letters of Intent* to Provide Financial Assistance to:

Peat Methanol Associates (First Colony Project), December 13, 1982.

Foster Wheeler Corporation and the Solv-ex Corporation (Santa Rosa Project), January 17, 1983.

Cool Water Coal Gasification Program, April 13, 1983.

Alberta Oil Sands Technology and Research Authority, Dynalectron Corporation, the Ralph M. Parsons Company, and Tenneco Oil Company (Calsyn Project), May 20, 1983.

Cathedral Bluffs Shale Oil Company, July 28, 1983.

Union Oil Company of California (Parachute Creek), December 1, 1983.

Ladd Petroleum Corporation, a subsidiary of the General Electric Company

HOP Kern River Commercial Development Project), February 1983.

The Dow Chemical Company (Dow Syngas Project), February 16, 1984.

Forest Hill Company and Greenwich Oil Corporation, April 5, 1984.

Texas Gas Development Corporation (Kentucky Tar Sand Project), April 5, 1984.

Northern Peat Energy Project, a wholly-owned subsidiary of Signal Energy Systems, April 5, 1984.

Gilbert Constructors, Inc. and Geokinetics, Inc. (Seep Ridge Project), June 22, 1984.

Great Plains Gasification Associates, consisting of Tenneco SNG, Inc., ANR Gasification Properties Company, Transco Coal Gas Company, MCN Coal Gasification Company, and Pacific Synthetic Fuels Company, April 26, 1984.

United States Synthetic Fuels Corporation Memorandum, "$92 per Barrel Briefing Papers", August 23, 1984. Author's files.

U.S. Synthetic Fuels Corporation, News Releases: June 2, 1982; June 18, 1982; January 20, 1983; February 17, 1983; March 16, 1983; April 13, 1983; May 27, 1983; June 30, 1983; October 22, 1983; December 1, 1983;February 16, 1984; April 5, 1984; April 26, 1984; June 18, 1984; July 2, 1984; October 22, 1984; March 20, 1985; April 23, 1985; May 22, 1985; May 29, 1985; June 19, 1985; June 28, 1985; July 31, 1985; August 22, 1985; September 11, 1985; September 25, 1985; October 17, 1985; November 20, 1985; and January 22, 1986.

United States Synthetic Fuels Corporation and Cool Water Coal Gasification Program, *Price Guarantee Commitment*, July 28, 1983.

United States Synthetic Fuels Corporation, the Dow Chemical Company, and Louisiana Gasification Technology, Inc., *Price Guarantee Commitment*, April 26, 1984.

United States Synthetic Fuels Corporation, Forest Hill Company. And Greenwich Oil Corporation, *Commitment to Guarantee*, September 24, 1985.

United States Synthetic Fuels Corporation and Union Oil Company of California, *Amendment Agreement*, October 16, 1985.

U.S. Synthetic Fuels Corporation, Solicitations for Proposals:

Initial Solicitation, November 21, 1980.

Second General Solicitation, December 11, 1981.

Third General Solicitation, August 19, 1982.

Fourth General Solicitation, February 16, 1984.

Competitive Solicitation for Western Oil Shale Projects, January 20, 1983.

Competitive Solicitation for Gulf Province Lignite Projects, April 25, 1983.

Competitive Solicitation for Eastern Province and Eastern Region of the Interior Province Bituminous Coal Gasification Projects, June 30, 1983.

Solicitation for Coal-Water Fuel Projects, January 5, 1984.

Solicitation for Coal or Lignite Gasification Projects, January 5, 1984.

Solicitation for Projects Proposing to Retrofit Existing Chemical Plants, February 16, 1984.

Solicitation for Eastern Province or eastern Region of the Interior Province Bituminous Coal Gasification Projects, May 28, 1985.

Solicitation for Projects to Produce Synthetic Fuels by Mining and Surface Processing of Tar Sands, June 24, 1985.

U.S. Synthetic Fuels Corporation, "Solicitation History", May 26, 1984. Internal SFC document, author's files.

U.S. Synthetic Fuels Corporation, Information Management, "Time Line", July 17, 1985. (Legislative History; Congressional and Other Legislative Activity; Appointments, Nominations and Resignations; Official Public Meetings; and Project and Other Corporate Milestones.) Internal SFC document, author's files.

U.S. Department of Energy, "Liquids from Coal", DOE/FE-0008.

U.S. Treasury Departmental Offices, "Office of Synthetic Fuels Projects Financial Status of Termination Activities for the Period Ending November 30, 1995", with an adjustment made by the author to account for the next three months' expenses when the monitoring office completed its work.

The Wall Street Journal, Editorial "Environmental Balance," July 13, 1990, New York Edition.

____. Editorial "Abuse of Power", March 22, 1989, New York Edition.

The Washington Post, "Casey Tied to Energy Firm's Bid for U.S. Backing," March 31, 1982.

____. "Suspect Firm a Finalist for Synfuels Aid," January, 1982.

____. "Synfuels Guarantee Put at $92 a Barrel", July 24, 1984.

U.S. Energy Research and Development Administration, *A National Plan for Energy Research, Development & Demonstration: Creating Energy Choices for the Future* (ERDA-40)

United States Synthetic Fuels Corporation, *Comprehensive Strategy Report*, June 1985.

United States Synthetic Fuels Corporation, *Comprehensive Strategy Report*, Appendices, "Table H-1. Assumed Characteristics of Pioneer and Second and Third Generation Plants", June 1985.

U.S. Synthetic Fuels Corporation Comptroller's Department, "Administrative Expenses History Authorization Use", October 25, 1985.

Notes

Chapter 1

1   U.S. Energy Research and Development Administration, *A National Plan for Energy Research, Development & Demonstration: Creating Energy Choices for the Future* (ERDA-40), p. II-5.

2   Sam H. Schurr and Bruce C. Netschert, et.al., *Energy in the American Economy, 1850-1975, Its History and Prospects* (Baltimore: The Johns Hopkins Press, 1960), 84

3   Ibid, 89

4   William Manchester, *The Last Lion: Winston Spencer Churchill ALONE 1932-1940* (Boston: Little, Brown, and Company, 1988), 438

5   U.S. Energy Information Administration, 9/30/2008.

6   Ruth Sheldon Knowles, *America's Energy Famine: Its Cause and Cure* (Norman: University of Oklahoma Press, 1980), 26

7   Ibid., 52

8   Ibid., 94

9   Ibid.

10   Ibid.

11   Ibid.,100

12   Ibid.,109

13   Ibid.,114

14   Ibid.,120

15   Lester A. Sobel (ed.), *Energy Crisis: Volume I, 1969—73* (New York: Facts on File, Inc., 1974), p.199.

16   Ibid., 178.

17   Stobaugh, Robert. *Energy Future* (New York: Ballantine Books, 1980), 18

18   Knowles, *America's Energy Famine*, 229.

19   Ibid.

20   Churchill, Winston. *The Gathering Storm*. New York: Houghton Mifflin, 1986. P.8.

21   Knowles, *America's Energy Famine*, 61

22   Ibid., 62

23   Ibid.

24   Ibid.,172

25   While a course in Economics 101 would be the appropriate place to impart knowledge of the workings of supply and demand in the market place, three fundamental conclusions are easy to grasp and pertinent to the above discussion: (1) supply will equal demand at some 'clearing price', (2) as prices rise, demand diminishes and supply is encouraged, and (3) conversely, as prices fall, demand is encouraged and supply discouraged. One might add to that, in any large commodity traded worldwide and involving multiple producers, there is little prospect of success for the market to be manipulated for any length of time.

26   Ibid., 166

27   Ibid., 208

28   Stobaugh, *Energy Future*,18

29   Ibid., 202

30   Ibid., 199

31   Ibid., 203

32   Ibid., 207

33   *Energy Crisis: Volume 3, 1975-77*, 35

34   Knowles, *America's Energy Famine*, 211

35   *Energy Crisis: Volume 3*, 37-44

36   *Energy Crisis: Volume 4*, 36-37

37   Knowles, *America's Energy Famine*, 219

38   Ibid., p. 221

39   *Energy Crisis: Volume 4*, 171

40   Ibid., 66

41   Knowles, *America's Energy Famine*, 234

42   *Energy Crisis: Volume 4*, 60

43   Ibid., 53

44   Ibid.

45   Knowles, *America's Energy Famine*, 231

46   Ibid., 235-236

Chapter 2

47   Wikipedia, Synthetic Fuels.

48   U.S. Department of Energy, "Liquids from Coal", DOE/FE-0008.

49   Ibid.

50   U.S. Energy Information Administration website, "South African Energy Data".

51   This section relies on the work contained in United States Synthetic Fuels Corporation, *Comprehensive Strategy Report*, June 1985, Appendix J submitted by the U.S. Synthetic Fuels Corporation to the Congress. While the numbers might have changed a bit the passage of time, the changes would not be material.

52   United States Synthetic Fuels Corporation, *Comprehensive Strategy Report*, June 1985, Appendix J.

53   Ibid.

54   Moreover, there are at least half a dozen characteristics of coal behavior that can vary considerably within one of these major designations and which can have substantial impact on a given coal conversion technology - i.e.: **Heating Value** (lower heating value coals require more coal to be mined and processed to make the same amount of synthetic fuel as higher heating value coals); **Reactivity** (lower reactivity coals require higher reactor temperatures and/or more time in the reactor to obtain acceptable conversion of coal to synthetic fuel); **Fines Content** (some synthetic fuels technologies are very sensitive to fines (particles less than ¼") and some coals have greater propensity to form fines during mining and handling); **Ash Melting Point** (some technologies perform more effectively with coals having high ash melting points; others with coals having low ash melting points); **Free Swelling Index** (some technology's performance are sensitive to coals with high free swelling indices); and **Sulfur Content** higher sulfur content coals may be more corrosive in, and immediately downstream from, the reactor and may cause higher pollution control costs).

55   In indirect conversion (gasification), the coal is first reacted with steam and oxygen to yield gas mixtures of CO, $CO2$, $CH4$, and $H2$, under conditions of moderate pressures (0-1,000 psi) and high temperatures (1,700-3,000 degrees F). United States Synthetic Fuels Corporation, *Comprehensive Strategy Report*, June 1985, Appendix J.

56   "... In a direct liquefaction (hydrogenation) system, hydrogenation takes place at moderate temperatures (700-900 degrees F) and in a high-pressure (2,000-4,500 psi) hydrogen environment. The elevated temperature causes the coal molecules to fragment, and the fragments react with hydrogen before they can react with each other (to form coke). The result is products with lower molecular weight, reduced oxygen, nitrogen, and sulfur levels, and higher hydrogen-to-carbon ratios." United States Synthetic Fuels Corporation, *Comprehensive Strategy Report*, June 1985, Appendix J.

Chapter 3

57   *Energy Crisis: Volume 4 1977-79*, pp. 39-41
58   "Compilation of the Energy Security Act of 1980, and the 1980 Amendments to the Defense Production Act of 1950", Chapter 6
59   Paper by Geoffrey A. McDermott, "The Energy Security Act of 1980", July 6, 1987, P.11
60   Ibid., p.10
61   Ibid., p.15
62   Ibid., p.18
63   Ibid.

Chapter 6

64   Edward W. Merrow, Kenneth E. Phillips, and Christopher W Myers, *Understanding Cost Growth and Performance Shortfalls in Pioneer Process Plants* (Rand Corporation, September 1981).
65   Ibid, p.4
66   Rand left the ultimate choice of projects to be included in the Study up to the participating companies. The companies, however, were requested by Rand to select "Plants that involved some degree or kind of technical change from prior plants, e.g., new process steps, new equipment, large scale-up, new plant configuration, etc. Additional selection criteria specified:
Medium-sized to large plants in terms of annual output—100 million pounds per year or more.
Plants that involve liquid and/or solids processing rather than strictly gas or cryogenic processes.
Plants constructed in the U.S. and Canada within the past 15 years.
Green-field, co-located, or add-on units but not revamps of an existing plant.
Plants for which reliable data are available.
No plant chosen solely because significant deviations from cost or expected performance occurred."
67   Rand Study, p.37
68   Ibid, p.47
69   Ibid, p.49
70   Ibid, p.67
71   U. S. Synthetic Fuels Corporation, *Annual Report 1981*.
72   Ibid.

Chapter 7

73 U.S. Synthetic Fuels Corporation, *Annual Report 1981.*

74 Ibid.

75 U.S. Synthetic Fuels Corporation *Annual Report 1982.*

76 Ibid.

77 Ibid.

78 Ibid.

79 Ibid.

80 Ibid.

81   U.S. General Accounting Office, *Circumstances Surrounding the First Colony Peat-to-Methanol Project,* November 20, 1983.

82 Letter of Intent signed by the U.S. Synthetic Fuels Corporation and Peat Methanol Associates (First Colony Project), December 13, 1982. Copies of all the letters of intent, which can be found in the U.S. Government General Records Center, were taken from the author's files.

83 This was the only agreement to provide for price guarantees lasting longer than ten years. To avoid the potential problems of such negotiations, all other agreements were negotiated on the basis of price guarantee payments being completed within the ten year period.

84 The proposed formula for sharing was for 50 percent of the excess over the Support Price to go to an Escrow Account or the SFC, and 50 percent to go to the project. Should market prices thereupon have fallen again below the Support Price, then subsequent price guarantee payments would be paid from the Escrow Account first, and, only when the account was exhausted, would the SFC again make payments. Once the price guarantee period expired, then all the funds in the Account and all subsequent revenue-sharing payments would go directly to the SFC. The project's obligation to share revenue, however, would cease when revenue-sharing payments equaled $210 million or equaled a lesser amount of total price guarantee payments made to the project.

85 Letter of Intent signed by the U.S. Synthetic Fuels Corporation, and Foster Wheeler Corporation and the Solv-ex Corporation (Santa Rosa Project), January 17, 1983.

86 The loan guarantee was determined on the basis of project costs being estimated at $32 million. Eligible costs were to include construction costs of the defined facility, interest paid during construction, commissioning and startup of the facility, and as much of Foster Wheeler's engineering costs as can be established as engineering for the facility up to a maximum of $440,000

87 The partnership's obligation to make cash flow payments was limited to amounts not exceeding one-third of the amounts distributed to

the partners during the first two-year period. And, aggregate Cash Flow Payments would not exceed $41 million.

88   Letter of Intent signed by the U.S. Synthetic Fuels Corporation and parties proposing to form a joint venture—the Alberta Oil Sands Technology and Research Authority, Dynalectron Corporation, the Ralph M. Parsons Company, and Tenneco Oil Company, May 20, 1983.

89   United States Synthetic Fuels Corporation and Cool Water Coal Gasification Program, *Price Guarantee Commitment*, July 28, 1983.

90   In the event (see Chapter 14), because of the plant's excellent operation in terms of high capacity factor and because inflation turned out to be less than that anticipated when determining the amount of obligational authority that would be required by the plant, the 20 trillion Btu limit proved to be limiting during the Project's fifth year of operation.

Chapter 8

91   U.S. Synthetic Fuels Corporation, *Annual Report 1983*.
92   Ibid.
93   Ibid.
94   U.S. Synthetic Fuels Corporation, *Annual Report 1984*.
95   Ibid.
96   Ibid.
97   U.S. Synthetic Fuels Corporation, *Annual Report 1985*.
98   Ibid.

Chapter 9

99   U.S. Synthetic Fuels Corporation, *Annual Report 1983*.
100 Ibid.
101 Where the narrative sometimes specifies 40,000 BOED capacity and sometimes 42,000 barrels per day of hydro treated oil, the difference is largely attributable to the fact that the crude shale oil swells as a result of the hydrogenation in the upgrading process.
102 Letter of Intent signed by the U.S. Synthetic Fuels Corporation and Union Oil Company of California (Parachute Creek), December 1, 1983.
103 The reduced levels of assistance for the equal-sized second increment shows the effect of economies of scale on projects of this kind.
104 Letter of Intent signed by the U.S. Synthetic Fuels Corporation and Cathedral Bluffs Shale Oil Company, July 28, 1983.
105 Letter of Intent signed by the U.S. Synthetic Fuels Corporation and Gilbert Constructors, Inc. and Geokinetics, Inc. (Seep Ridge Project), June 22, 1984.

106 As will be discussed in Chapter 11, the amounts of assistance were renegotiated substantially upward in 1985 to reflect changing circumstances of the market.

107 Letter of Intent signed by the U.S. Synthetic Fuels Corporation, and Forest Hill Company and Greenwich Oil Corporation, April 5, 1984.

108 Letter of Intent signed by the U.S. Synthetic Fuels Corporation and Ladd Petroleum Corporation, a subsidiary of the General Electric Company (HOP Kern River Commercial Development Project), February 1983.

109 Inasmuch as the project was estimated to cost $151.9 million, the rest of the financing was to include $42.4 million of contributed capital and $44.3 million of internal cash generation. As such, the guaranteed debt amounted to only 42.9 percent of the project's cost vice the maximum of 75 percent allowed by the Act.

110 Given the incremental nature of this project's intended development, and the perceived need by the Corporation to provide material incentives for completing the project, limitations were placed on the amounts of Product that would be eligible to receive price guarantee payments that were a function of the number of "Units" actually constructed.

111 Letter of Intent signed by the U.S. Synthetic Fuels Corporation and Texas Gas Development Corporation, April 5, 1984.

112 The proposed agreement also would have capped the amount of payments that could be made in various quarters to help ensure the longer term operation of the facility. Specifically, payments could not exceed $17 million in any of the first twelve quarters of the Plant's operation, and could not exceed $9.3 million in any subsequent quarter.

113 Letter of Intent signed by the U.S. Synthetic Fuels Corporation and Northern Peat Company, a wholly-owned subsidiary of Signal Energy Systems, Inc., itself a wholly-owned subsidiary of the Signal Companies, Inc., April 5, 1984.

114 Because the synthetic fuel was to be sold as a substitute for fuel oil in most instances, the market price to be used for the calculation of price guarantee payments was to be the New York posted price for residual No. 6 Fuel, 2.2 % sulfur. The actual formula was still to be determined between the sponsor and the Corporation when preparing the final contract so as to take into account actual pricing clauses in sales contracts as well as the effects of spot market pricing factors.

115 Letter of Intent signed by the U.S. Synthetic Fuels Corporation and the Dow Chemical Company, February 16, 1984.

116 United States Synthetic Fuels Corporation, the Dow Chemical Company, and Louisiana Gasification Technology, Inc., *Price Guarantee Commitment*, April 26, 1984.

117 U.S. Synthetic Fuels Corporation, *Annual Report 1985*.

118 By employing a mix of indices, the negotiators had attempted to match the Project's likely increase of costs (as opposed, for example, to tracking the average increase in consumer or producer prices).

119 These restrictions were to constrain plant operations in it last period of operation and were renegotiated as discussed in Chapter 14.

120 The Great Plains Coal Gasification Project was funded by DOE under the preliminary program and was intended to be transferred to the SFC as had Parachute Creek and the Tosco project. Political considerations caused its administration to remain with the DOE.

Chapter 10

121 U.S. Energy Administration, "Annual U.S. Crude Oil Imports from All Countries", http://tonto.eia.gov., October 15, 2008.

122 At the time of this writing in 2008, import levels had increased to 12 million barrels a day, two-thirds of domestic consumption, and prices spiked to $147 per barrel ($57 per barrel in 1985 dollars).

123 Public Law 98-473, "Making Continuing Appropriations for the Fiscal Year 1985, and for Other Purposes."

124 United States Synthetic Fuels Corporation, *Comprehensive Strategy Report*, June 1985.

125 United States Synthetic Fuels Corporation, *Comprehensive Strategy Report*, June 1985, p. 33.

126 Ibid., p. 38.

127 United States Synthetic Fuels Corporation, *Comprehensive Strategy Report*, Appendices, "Table H-1. Assumed Characteristics of Pioneer and Second and Third Generation Plants", p. H-10, June 1985.

128 Hagler, Bailly and Co. *Learning Curves: Empirical Evidence of Production Cost Improvement*, HBC Ref. No. RA85-420-2; Washington, D.C., June 1985.

129 R.W. Hess. *Potential Production Cost Benefit of Constructing and Operating First-of-a-Kind Synthetic Fuel Plants*, WD-2577-SFC. The Rand Corporation, March, 1985.

130 United States Synthetic Fuels Corporation, *Comprehensive Strategy Report*, Appendices, "Table H-1. Assumed Characteristics of Pioneer and Second and Third Generation Plants", p. H-15, June 1985.

131 Ibid., p. H-17.

Chapter 11

132 The $327 million loan guarantee amount was derived from a cost estimate of $560 million to complete the Augmentation Program. That total included not only an amount of $286 million to incorporate the additional facilities associated with the fluidized-bed combustion technology, but also the costs of maintaining the existing facility until the Augmentation facilities would allow the entire plant to come into operation—in the 1989 - 1990 time frame. The guaranteed loan would amount to only 55 percent of the total costs with Unocal providing the remaining amount in equity. (The Act permitted the Corporation to guarantee up to 75 percent of the costs of a facility.)

133 This change had no impact on the amount of subsidy that would be provided by the Corporation since it was paying the difference between market and guarantee prices. The difference between $37.87 and the market price for upgraded shale oil equivalent, and $42.50 and the market prices for diesel and jet fuels would be about the same.

134 United States Synthetic Fuels Corporation and Union Oil Company of California, *Amendment Agreement*, October 16, 1985.

135 United States Synthetic Fuels Corporation, Forest Hill Company. And Greenwich Oil Corporation, *Commitment to Guarantee*, September 24, 1985.

136 Letter of Intent signed by the U.S. Synthetic Fuels Corporation and Great Plains Gasification Associates, consisting of Tenneco SNG, Inc., ANR Gasification Properties Company, Transco Coal Gas Company, MCN Coal Gasification Company, and Pacific Synthetic Fuels Company, April 26, 1984.

137 Frank Herrington, Secretary of Energy, letter to Edward E. Noble, Chairman of the Board SFC, May 21, 1985.

138 United States Synthetic Fuels Corporation, *News Release*, July 2, 1984.

139 United States Synthetic Fuels Corporation, *1985 Annual Report*.

Chapter 12

140 Ironically, as seen in Chapter 10, he was to be appointed to the new Board of the SFC late in 1984.

141 Schumacher, E.F. *Small is Beuatiful*. 1973.

142 Hawken, Paul, Ogilvey, James, and Schwartz, Peter. *Seven Tomorrows*. 1982

143 U.S. Senate, Committee on Government Affairs/ Subcommittee on Oversight of Government Management. Hearing on SFC Management.

98th Cong., 1ˢᵗ Sess., July 27, 1983, and July 29, 1983.

144 U.S. House of Representatives, Committee on Energy and Commerce/ Subcommittee on Oversight and Investigations. Hearings regarding SFC President Victor Thompson. 98th Cong., 2ⁿᵈ Sess., April 3, 1984.

145 U.S. Synthetic Fuels Corporation Legal Services Group, Memorandum for the Board of Directors, "Victor M. Thompson, Jr./ Utica Bankshares Corporation, April 23, 1984.

146 The author as Vice President for Projects was present.

147 V.M. Thompson, Jr., Letter to the Board of Directors, United States Synthetic Fuels Corporation, April 26, 1984.

148 U.S. Synthetic Fuels Corporation, Analysis of Administrative Issues, April 24, 1985, Author's Files.

149 U.S. Synthetic Fuels Corporation, Analysis, September 17, 1985.

150 U.S. Synthetic Fuels Corporation, Analysis, September 17, 1985.

151 Also, Synar was a Democrat from Oklahoma, a state from which Noble had tried to run for the Senate as a Republican.

152 The author attended virtually all of the congressional hearings at which the Corporation testified, either at the witness table with the principal SFC witness or at the side as a resource. In addition, he and other senior members of the Corporation, provided briefings for the staffs of about five of the subcommittees on almost a monthly basis over a several year period, keeping them fully abreast of the Corporation's activities and the rationale for its actions.

153 U.S. House of Representative, Committee on Government Operations/ Subcommittee on Environment, Energy and Natural Resources. Hearing regarding an "Examination of Procedures by which the SFC Selects Projects for Federal Financial Assistance". 98th Cong., 1ˢᵗ Sess., December 8, 1983.

154 U.S. House of Representatives, Committee on Government Operations/ Subcommittee on Environment, Energy and Natural Resources. Hearing on the proposed award of assistance to the Great Plains Project. 99th Cong., 1st Sess., May 22, 1985.

155 U.S. House of Representatives, Subcommittee on Oversight and Investigations of the Committee on Energy and Commerce, Letter from John D. Dingell and James T. Broyhill to Edward E. Noble, February 16, 1984.

156 Jeffrey Lipkin (previously an SFC General Counsel) E-mail to Hervey Priddy, February 24, 2011.

157 The author conducted the press conferences for the first two years or so of the Corporation's existence; thereafter Tom Corcoran did so when he became Vice Chairman.

158 In contrast, Martha Hamilton wrote an extensive and well re-

searched piece on the outcome of the four projects funded by the SFC (*The Washington Post*, June 26, 1988). This, however, was published long after the political battle was over.

159 U.S. Synthetic Fuels Corporation, Letter from Edward E. Noble to the Editor of the *Wall Street Journal*, October 9, 1985.

160 Congress of the United States, House of Representatives, Letter from Member of Congress Michael L. Strang to the Editor of the *Wall Street Journal*, October 18, 1985.

161 William F. Rhatican, Letter to Jack Anderson, August 25, 1983.

Chapter 13

162 *U.S. Congressional Record - House*, July 25, 1984.

163 Ibid.

164 For every piece of legislation, the Rules Committee structures how the bill will be handled on the floor of the House, from the allotment of debate time to a list of which amendments can be offered. Given that the Committee is controlled by the Speaker, it is very rare that the House votes to overturn the Committee's actions. In this case, the House leadership did not want a waiver of Clause 6 of rule XXI, which prohibited a general appropriation bill from containing a reappropriation of previously appropriated funds.

165 *U.S. Congressional Record - House*, July 25, 1984.

166 Even though Dingell was one of the leading opponents of the SFC, as a member of the Democratic hierarchy, he was necessarily against a waiver of the rule, which was the subject of debate.

167 United States Synthetic Fuels Corporation, *1985 Annual Report*.

168 United States Synthetic Fuels Corporation Memorandum, "$92 per Barrel Briefing Papers", August 23, 1984.

169 *U.S. Congressional Record — Senate*, September 26, 1984.

170 228. HR3011. Interior Department Appropriations, Fiscal 1986. July 24, 1985.

171 255. HR 3011. Interior Department Appropriations, Fiscal 1986. July 31,1985.

172 *U.S. Congressional Record - Senate*, October 31, 1985.

173 United States Senate, Committee on Energy and Natural Resources, Letter from the Chairman, James A. McClure to Colleagues, November 7, 1985.

174 Executive Office of the President, Office of Management and Budget, Letter from James C. Miller III, Director, to the Honorable Mark O. Hatfield, United States Senate, December 12, 1985.

Chapter 14

175 United States Synthetic Fuels Corporation, *1985 Annual Report*.

176 Dakota Gasification Company Internet Site (9/24/08)

177 United States Synthetic Fuels Corporation, *1985 Annual Report*.

178 SFC Annual Reports for 1981, 1982, 1983, 1984, and 1985.

179 U.S. Treasury Departmental Offices, "Office of Synthetic Fuels Projects Financial Status of Termination Activities for the Period Ending November 30, 1995", with an adjustment made by the author to account for the next three months' expenses when the monitoring office completed its work.

180 Energy Security Act, Section 100(b)(2))

181 Electric Power Research Institute, *Cool Water Coal Gasification Program: Final Report* (EPRI GS-6806, Project 1459), December 1990.

Appendix A

182 Source: Table J-12 United States Synthetic Fuels Corporation, *Comprehensive Strategy Report*, June 1985, Appendix J.

Appendix D

183 "Learning Curves: Empirical Evidence of Production Cost Improvement," HBC Ref. No. RA85-420-2; Washington, D.C., June 1985.

Index

www.ingramcontent.com/pod-product-compliance
Lightning Source LLC
Chambersburg PA
CBHW021547210326
41599CB00010B/339